DECEIVED WISDOM

Deceived Wisdom

Why what you thought was right is wrong

DAVID BRADLEY

First published 2012 by
Elliott and Thompson Limited
27 John Street, London WC1N 2BX
www.eandtbooks.com

ISBN: 978-1-908739-34-6

Design and illustration by Louis Mackay / www.louismackaydesign.co.uk

Printed by TJ International Ltd.

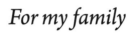

For my family

Contents

. . . .

Introduction

. . . .

Did your mother remind you to take off your coat when you came inside, or you wouldn't 'feel the benefit' when you went out again? Have you ever been informed that what you need to cool down is a nice cup of tea? And are you bored with being told that you have to give red wine time to breathe to improve its taste? If so, then *Deceived Wisdom* is the book for you. It looks at things we think we know and examines why we don't know them at all.

There is much deceived wisdom in the world – from fitness fallacies to dietary deceptions and countless miscellaneous misconceptions. Given that human beings are inquisitive by nature, it might seem odd that there is so much misinformation swirling around and confusing us, but perhaps it's not that surprising. We have come to expect definitive answers to our questions in an instant rather than having to puzzle over a problem and work it out for ourselves, and it's tempting to reach for an instant answer without worrying about whether it's right or not. It's easier to remember a piece of received wisdom – say a memorable aphorism or a sound bite – than to look more closely and extract the truth.

Yet to accept without question any statement that comes your way is to miss out on a world of fascinating facts – and all too often the received wisdom is simply incorrect. So don't let your view of the world be based on hearsay or old wives' tales or whatever you want to call them. I want to persuade you to embrace a science-based approach, to weigh up the evidence and come to a scientific conclusion about reality. Science can use Occam's razor to shave down overly complex explanations to the essential facts: the simplest explanation, rather than the most convoluted, will usually suffice.

So if you want to get your facts straight, read *Deceived Wisdom* and be ready to put the know-alls in their place and debunk their misbegotten myths.

Can a cup of hot tea help you break the law?

The Deceived Wisdom
A nice hot cup of tea on a warm day
cools you down.

• • • • • • • • • • • •

As climate change kicks in and global temperatures rise, there is a disturbing trend to invoke spurious advice on keeping cool. Some of that advice dates back to the time of Victorian prime minister William Gladstone. Gladstone is perhaps better known for his six decades in British politics and his efforts to rehabilitate London's prostitutes than for his advice on human temperature control.

But is there a nugget of truth in Gladstone's claim that a hot cup of tea can cool you down on a warm summer's day? It may be that this piece of deceived wisdom was nothing more than an excuse for the British to drink tea on any occasion regardless of the weather. After all, where such received wisdom relies on the hearsay and false truths of old wives and their elderly husbands, deceived wisdom has science to which it can turn to debunk the perpetrators of the misconceptions.

The first law of thermodynamics tells us that adding heat to a system will make it hotter. It seems so obvious, but it was not until the mid 1800s that physicists laid the foundations of our modern understanding of thermodynamics and helped put the steam-driven technology of the Victorian era on a firm scientific footing.

Physiologically, things might not be so clear-cut. The human body has an internal feedback system that usually keeps the blood from overheating and the internal organs stable. Drink a hot drink, and yes, the temperature of your stomach's contents will rise, but that will also bring about a slight hastening of the heart, an expansion of the blood vessels close to the surface of the skin, and an increase in sweating as the brain switches on the various feedback-controlled temperature regulators to maintain the body at its normal temperature of about 37°C.

It is that word 'feedback' that provides a clue as to why a cup of hot tea gained its reputation as an effective cooling agent. Feedback loops always have a time lag. So the instant burst of warming that comes from sipping a nice hot cup of tea will inevitably bring you out in a bit of a sweat on a hot day as the brain fights to compensate for the localized rise in temperature in your stomach. The compensatory measures take time to be reversed once the normal temperature balance is restored, and their effects might last slightly longer than the temperature regulation process needs. However, there is no escaping the long arm of the first law: adding the hot liquid to your cooler stomach raises its temperature. Your skin may feel slightly cooler because of the evaporative cooling effect of sweating, but your body temperature will quickly return to that average 37°C.

We are more than vessels for receiving hot tea, of course. Perhaps the real reason that old wives and elderly husbands believed that a hot cup of tea cools you down was more to do with interrupting whatever activity was making you hot in the first place. If you have abandoned the mad dogs and Englishmen out in the midday sun, then you will most likely have stepped indoors, filled the kettle, and settled down to the ceremonious act of making and drinking a pot of tea. The whole process of tea-drinking is often relaxing, and frequently refreshing, but thermodynamically never cooling.

There is another thermodynamic consequence that puts paid to the deceived wisdom that 'you will not feel the benefit' if you keep your coat on indoors before going out into the cold once more. Coats are usually designed as insulators – a coat works by trapping air within the tiny spaces between its fibres and between it and the layer of clothing underneath. Keeping your coat on will ensure that less body heat is lost and that the air between those fibres is kept warm. Warm air is a better insulator than cold air.

So when you head outside again with your coat still on, you will be warmer than if you had removed your coat. The only proviso comes with that concept of the body's internal temperature control. If you get too sweaty indoors with your coat on, then evaporative cooling might make your skin temperature drop when you step outside, so you may well not feel the benefit, but instead feel the rush of cold air whipping away your body heat.

* * * * * * * * * * * *

The Science
Adding a hot liquid (a cup of tea at 50 to 60°C) to a cooler vessel (your stomach at 37°C) raises your stomach's temperature. This slight increase in body core temperature may well

cause the brain to stimulate increased sweating
to counteract this rise in temperature, but within
minutes your body will be back to its normal
temperature of 37°C.

• • • • • • • • • • • •

Find out more

For: http://coffeetea.about.com/cs/whimsy/a/teamyth.htm
Against: http://www.sennir.co.uk/Journal/Does_Tea_Cool_You_Down

Cracking passwords

The Deceived Wisdom

Even a seemingly random mix of numbers, symbols and upper- and lower-case letters does not make a perfectly uncrackable password, despite what the online password-strength meters might suggest.

• • • • • • • • • • • •

A password is a key. A key that allows you to lock up something you consider important or otherwise want to keep secret. In ancient times a password might allow you to pass through the city gates after hours; in spy thrillers it can convince the double agent you are hoping to bring in from the cold that your credentials are valid.

In computing, a password is a string of characters – letters, numbers and symbols – that is understandable (and ideally memorable) to the individual. It is used to encrypt data so that the data cannot be read by anyone who does not have the password. Without passwords and encryption, there would be no security when you log into your email, do your internet shopping or check your credit card statement online.

Encryption involves the superficially simple process of

transforming the readable stream of data, using a computer program or algorithm – the cipher – into a new data stream that is unreadable to another computer without the key – the password – to that cipher. Strong passwords and strong encryption algorithms are vital for safeguarding our finances during online transactions, and even for seemingly minor things such as Twitter updates. Unfortunately, there are always those who would like to steal their way past the guards and pick the locks or crack the passwords, either for personal gain or out of simple malice.

What is the best kind of password to keep your data protected? Obviously it should be one that keeps your login secure and is not going to be cracked. There are several schools of thought on what constitutes a good, strong password. Sites that test the strength of your password will have specific criteria for deciding what they consider strong: password length, mix of upper- and lower-case letters, numbers or characters including duplicate letters, and so on, and so may give you a false sense of security depending on how they are set up to test your choice.

The first approach is to create a long 'random' string of letters (upper and lower case), numbers and characters. Tools such as LastPass, KeePass and other password-storage programs can generate such strings for you based on different criteria. For example, this is a password generated by KeePass: Jc\z'ofg5^fhr951x.`eUTHDaO. I set the program to allow upper- and lower-case letters, numbers and other symbols from the computer keyboard. (Obviously, I don't plan to use this password for my credit card login, so don't bother trying.) Such passwords are almost impossible to remember without a program to use as a password locker (an application that itself can be password-protected to store passwords for other sites in an encrypted format). But if you go down this route, how do you set and remember the password for your password locker?

I used one of the many online password meters[1] to test this generated password. It tells me that the password is '100 per cent Very Strong' based on the mix of characters and the length of the password. The password tester from software giant Microsoft[2] categorizes this password as 'Best'. Another test claims that it would take a desktop computer about 438 decillion years (that is 438 followed by 33 zeros – slightly longer than the age of the known universe, you might say) to crack it. So it seems that Jc\z'ofg5^fhr951x.`eUTHDaO would indeed make a secure password: one that, though hard to remember, would be very hard to crack.

Many websites, even those of some banks and other financial services providers, do not allow such long or complex passwords and force you to devise a password that contains only alphanumeric characters. They often require or exclude numbers and sometimes restrict you to a small number of characters. This is dangerous. A password just eight characters long consisting of a random string of letters could theoretically be cracked quickly given a powerful enough computer, or network of computers, and a truly dedicated cracker.

Cryptographers talk of 'password entropy' – a term borrowed from the physical sciences. Entropy is a measure of disorder. A crystalline solid in which the atoms are arranged in regularly repeating patterns, like a microscopic, three-dimensional wallpaper print, has less entropy than the same material in the liquid state, in which the atoms are free to move

---

1   http://www.passwordmeter.com

2   http://www.microsoft.com/en-gb/security/pc-security/password-checker.aspx

randomly and there is no order or repeating pattern. Similarly, a password based on a random string of characters has more entropy than a dictionary word or a password like 'abababab'.

The entropy of a password is measured in bits and is a measure of its strength, based on the number of random guesses one would need to make to hit upon the actual password. A password with 32 entropy bits, where each bit has been picked randomly by the toss of a coin, would require 2 to the power 32 ($2 \times 2 \times 2 \times 2$ and so on, 32 times) tries before all possible combinations were exhausted. Adding one more bit ($2 \times 2 \times 2 \times 2$ and so on, 33 times) doubles the entropy, meaning that twice as many guesses would have to be tried before the random password was cracked. Of course, there is always the chance that a password cracker will guess right first time, while others will guess right only after trying all the other possible passwords.

There is a second approach to password creation that is gaining some credence among security experts. That is to create a password simply using four random words that you can learn easily for recall later. For example, you might pick 'sliver', 'finger', 'purple' and 'breakfast'. Your password would then be 'sliverfingerpurplebreakfast'.

This password does not seem to meet most of the criteria used by standard password tests. One of the online testing systems warns me that it looks like a word or a name. Of course, it is obviously not a real word, and if you pick a random combination of words of all types, perhaps from other languages, it is very unlikely that they are going to appear together in any dictionary or cracker list of passwords to try first, unlike 'password' and its ilk.

The test sites also flag up the fact that this password contains no non-alphanumeric characters. Be that as it may, the test also says that it will take a brute-force attack 20 sextillion years (that's a 2 with 22 zeroes after it) to crack the password just based on random guesses one after another. Of course, it is possible to make the cracking time longer by mixing in some upper-case letters and adding some numbers without making the password impossible to remember – 'sliverFingerPurple321breakfast', for instance.

No password is impossible to crack, but some take quite a few years longer to crack than others. To be even more secure, change your passwords frequently. Whatever you do, don't make your password 'password' or '123456', but do make sure you can remember it without resorting to writing it down somewhere an intruder might find it, like a sticky note attached to your monitor …

• • • • • • • • • • • • •

The Science
There is no perfect password; given enough time and computer power, there is always a way to crack a password. However, if your password is 'passwd1', '123456', your mother's maiden name, your wedding anniversary, a pet's name or something equally identifiable, then you are likely to be cracked sooner, rather than later.

• • • • • • • • • • • • •

Find out more
http://www.symantec.com/connect/articles/
ten-windows-password-myths

A different kind of snow

The Deceived Wisdom
No two snowflakes are alike.

• • • • • • • • • • • •

Generations of primary school children have attempted to simulate nature in their classrooms in the run-up to Christmas. They carefully cut out circles of white paper, and fold them into halves, quarters and even eighths. They eagerly snip away at the edges with safety-conscious round-ended scissors. Finally, they unfurl their paper to reveal beautiful eight-sided snowflakes with which to decorate the classroom. At the end of term, they retrieve their decorations and offer doting parents the opportunity to adorn kitchen noticeboards and windows around the home in celebration of ancient pagan and religious festivals.

Pedantic parents, when they accept the snowy offerings, might choose to point out that snowflakes (or more correctly, snow crystals) are never eight-sided. The chemistry inherent in water molecules ensures that all snow crystals form with hexagonal, six-sided symmetry. Those same pedantic parents may also point out that no two snow crystals are alike, which is a property shared with the paper simulations.

And, they would be right ... to a point. If we investigate the structure of those tiny snow crystals down to the microscopic, submicroscopic, and even atomic levels, then it would be impossible for any two snow crystals to be identical. To be identical, each one of countless molecules of water locked into each icy structure would have to be in exactly the same position in each and every crystal. A single snow crystal might weigh a milligram (a thousandth of a gram) and so contain a billion billion water molecules (that's 1,000,000,000,000,000,000). The possible variations between two snow crystals are to all intents and purposes infinite – and that's before we take into account that there is more than one type of oxygen atom (different isotopes of oxygen differ in the number of neutrons in their nucleus), so any one of the billion billion oxygen atoms in all those water molecules might be an oxygen-18 instead of an oxygen-16 and be in any position. The natural abundance of the heavier form of oxygen is about one in 500, so for every 500 oxygen atoms in a snow crystal one of them might be oxygen-18, and it could be any one of those 500. The chance of the same pattern existing in another snow crystal is much more remote than the chance of you winning the National Lottery.

As if that weren't enough, to see those kinds of differences requires extremely powerful microscopes. You would also need to gather up all the snow crystals and compare each one with the all of the others without damaging them and without any of them melting. It would be quicker to leave an infinite number of monkeys with an infinite number of typewriters to come up with 'To be, or not to be.'

But if you're simply taking a look at snow crystals with the naked eye, then you'll quite likely spot several that are indistinguishable. Indeed, apart from their shape and the substance from which they are made, snow crystals are not too different from crystals of table salt, and you would not expect to see much variation in those tiny little crystalline cubes.

As salt crystals grow with cubic symmetry, so snow crystals form with hexagonal symmetry. A tiny speck of dust acts as the nucleation point on which water vapour will condense from the atmosphere and form ice. The ice crystals grow with hexagonal symmetry because loose bonds, known as hydrogen bonds, can form between individual water molecules. These hydrogen bonds exist only fleetingly in the liquid but are locked in place in ice. Water molecules (H_2O, or H—O—H where the dashes represent the internal chemical bonds between the hydrogen atoms and the oxygen atom) are actually bent at an angle, and this forces the hydrogen bonds between the water molecules to configure themselves in a way that gives rise to an overall structure with an overall hexagonal symmetry – hence the six-sided or six-armed snow crystal. In nature, as opposed to primary school classroom walls, you will never see an octagonal or even pentagonal snow crystal. They are always hexagonal.

Why are snow crystals so important, and why do scientists care about them so much? Well, the formation of snow and ice crystals in the atmosphere can affect climate and the health of our planet. For instance, ice reflects light very well, so the amount of ice on the earth's surface determines how much energy from the sun is reflected back into space. Ice crystals can act as catalysts for the breakdown of ozone. Ozone is a form of oxygen found in the upper atmosphere that absorbs harmful ultraviolet rays from the sun and prevents them from reaching the earth's surface. Without the ozone layer, life as we know it could not survive. The presence of ice and snow crystals in the atmosphere also plays a critical role in the build-up of electric charges in clouds, which ultimately leads to lightning storms.

One question remains when considering the six-armed type of snow crystal, as opposed to the type that are more like hexagons with holes. How does each of those six arms 'know' to grow in the same branching way as the other five? How do all those countless water molecules 'know' to freeze into place to make the snow crystal perfectly symmetrical? They are growing under almost identical conditions, of course, so one might expect this to be the case. But it is only to the naked eye and on primary classroom walls that snow crystals are perfect – though not necessarily in the eyes of pedantic parents. Under the microscope, it is easy to see that six arms of that type of snow crystal are not all identical: there may be a tiny protuberance missing from one or an extra tiny knob of ice on another that make them deviate from perfection.

. . . .

The Science
To the naked eye, snow crystals look like just so many tiny hexagons or six-sided wheels, with a few variations on the theme. But get a lot closer and it becomes clear that no two snow crystals could ever be exactly the same.

. . . .

Find out more

http://www.its.caltech.edu/~atomic/snowcrystals/alike/alike.htm

Snack-sized dietary deceptions

• • • • • • • • • • • •

ACNE IS MORE THAN SKIN DEEP – For years now, parents have been warning teenagers to cut down on chocolate, sweets and greasy fast food so that they do not burst forth with acne. Acne vulgaris afflicts many pubescent youths and some adults. It has been referred to by at least one cynical and bitter observer as nature's contraceptive because of its often detrimental effect on one's appearance.

However, acne is not caused by overindulgence in fatty and sugary foods: it is simply a skin-deep response to the rapid rise in the sex hormone testosterone during puberty, though there is some evidence that a high glycaemic load (too much sugar in the blood) can make a case of acne worse. A tendency to get acne also runs in families, so you may be stuck with those spots no matter what you do. The advice that teenage offspring should tame their voracious appetite for sweet snacks and fast foods is still worth taking, though: it could help to reduce the risk of becoming overweight or developing serious long-term health problems such as type 2 diabetes and obesity.

• • • • • • • • • • • •

Find out more: http://en.wikipedia.org/wiki/Acne#Cause

"

Be careful about reading health books. You may die of a misprint

"

Mark Twain

EATING CARROTS WILL HELP YOU SEE IN THE DARK – There is a slice of truth to the dietary deception that noshing on the orange root vegetable will improve your vision. During the second world war, the British spread a rumour that their bomber pilots had perfect night vision because they were eating so many carrots. The rumour was allegedly part of the propaganda machine to prevent the Germans from discovering that the Allies had developed radar, which really can see in the dark.

Carrots and many other vegetables contain relatively large amounts of vitamin A, which also goes by the chemical name retinol. This compound is needed to produce the pigments in the retina at the back of the eye and so maintain healthy eyesight. However, only tiny amounts of all vitamins are necessary – and overindulging in certain ones, particularly the fat-soluble vitamins, can prove harmful. It is possible to overdose on some vitamins by ingesting an excessive amount of vitamin A from supplements. An excess of vitamin A in your diet can accumulate in your body, leading to toxic effects such as headaches, dizziness, vomiting, peeling skin, liver problems and, ironically, blurred vision.

· · · · · · · · · · · · ·

Find out more:

http://www.snopes.com/food/ingredient/carrots.asp

FEED A COLD, STARVE A FEVER – I am a big food fan. Only a really bad upset stomach will put me off my meat and two veg. Never when I've caught the common cold or had a dose of flu do I not feel like eating. This is just as well, because this particular bit of dietary deception is phoney.

High fevers and the common cold often cause fluid loss through sweating and increased urination. As such, you have to replace those lost fluids by drinking plenty of water, fruit or carrot juice, chicken soup or just a cup of warming tea. A large proportion of the water we get each day comes from moisture in our food. But when you are ill it is also important to get enough nutrition from your food: vitamins, minerals, salt and energy-giving sugars. Fever or cold, it is best not to starve – otherwise you will deprive your body of the essential nutrients it needs to fight the illness.

* * * * * * * * * * * *

Find out more

http://www.dukehealth.org/health_library/health_articles/feed_a_cold

IF FOOD DROPPED ON THE FLOOR IS THERE FOR FIVE SECONDS OR LESS, IT'S STILL SAFE TO EAT IT – Countless teenagers throwing caution to the wind and, naturally, ignoring parental advice, love to eat food like pizza, burgers and fries. And among teenagers there is the well-known opinion that food dropped on the ground and retrieved within five seconds is perfectly safe to eat, whether the pizza lands pan-side down or the cheese and tomato mix hits the pavement. Said in jest it may be, but there's no denying that people follow it.

Unfortunately, bacteria, viruses and the parasites present in animal faeces, which are so often smeared on pavements in town centres, can survive for long periods on dirty surfaces. Germs are much faster-moving than even the sportiest of teenager stoked up on fries and will clamber aboard any dropped food within much less than five seconds. When invoking the five-second rule, first think salmonella, toxicaria and *E. coli* before taking another bite.

• • • • • • • • • • • •

Find out more

http://www.clemson.edu/public/psatv/health/five-second-rule.html

LICKING THE BOWL WILL GIVE YOU WORMS – Cake-makers are often inefficient in scooping out all of the sticky mixture of flour, sugar and eggs from the bowl. This gives anyone hanging around the kitchen the chance to dip their finger into the bowl for a sweet and gloopy taster before the washing-up gets done. Generations of cake-makers have, however, warned such opportunists of the perils of eating the raw mixture – the worst of which is that the raw flour will give the finger-licker worms.

Worms are parasites whose rather unpleasant life cycles depend on the inadvertent transfer of eggs from faecal matter to your mouth. You can catch worms by swimming or bathing in contaminated water or by eating inadequately cooked meat. There is no mention of raw flour in the textbooks on parasitic worms. You might acquire a salmonella infection from the raw eggs though. Lingerers be warned: leave the bowl-licking to the cake-maker.

• • • • • • • • • • • •

Find out more

http://www.themangotimes.com/themt/2006/2/21/parental-myths-7-raw-dough-and-worms.html

The big cheese and one small step

The Deceived Wisdom
Conspiracy theories, speculation and rumour circulated on countless websites and by email claim that astronauts have never landed on the moon.

• • • • • • • • • • •

I f you are lucky enough to have clear skies on the night of a total lunar eclipse, you can watch in awe as the shadow of the earth gracefully crosses the surface of our planet's nearest neighbour. The moon – the great ball of 'cheese' in the sky, subject of song and sonata, poem and painting, and the focus of all those who have gazed at the night sky with wonder since ancient times.

The moon was always a target for humankind's exploratory urges. We simply had to make that giant leap to set foot on the moon, to find out what this other world is really like. The cold war posturing of the old Soviet Union and the United States culminated not in missile tests or nuclear demonstrations, but in the race to the moon. On 20 July 1969 the race reached its climax when the *Eagle* came home to roost, and NASA astronaut Neil Alden Armstrong stepped down from the Lunar Module of Apollo 11 and created the first extraterrestrial human footprint. That footprint remains, in the volcanic plain we know as the Sea of Tranquility, Mare Tranquillitatis.

When he made his indelible footprint, Armstrong uttered what became one of the most famous quotations in history. Such a momentous occasion called for momentous words, of course, and Armstrong may have rehearsed them again and again on the long journey from the Kennedy Space Center on Merritt Island, Florida – though he tells interviewers that he wrote them during the hours that passed while he and fellow astronaut Buzz Aldrin remained in the *Eagle* before finally, dramatically, opening the hatch.

'That's one small step for man, one ... giant leap for mankind.'

On hearing those crackly words, one can imagine his disappointment. He had just stepped on to the biggest and brightest stage in the history of humanity and ... well, to put it bluntly, he fluffed his one important line. Surely it should have been 'One small step for *a* man, one giant leap for mankind', without the pause, with the missing 'a'. That 'a' is critical, because without it the statement doesn't make sense. 'One small step for man' is synonymous with 'one small step for mankind', not with 'one small step for a man', the single, pioneering individual, mission commander Armstrong.

And what's with the pause? That almost imperceptible but implicitly infinite hesitation between Armstrong's 'one' and his 'giant' is the moment at which he realized the error of his words. His pioneering first step, the first small step by a giant of a man on to another world, would be indelibly marked for all eternity by a blooper.

> **"**
>
> **What we call rational grounds for our beliefs are often extremely irrational attempts to justify our instincts**
>
> **"**

Thomas Henry Huxley

The quotation, or shall we say misquotation, has caused consternation for over four decades. Some say the 'a' was lost in transmission through radio interference and static noise across the 400,000 kilometres between Armstrong's microphone and mission control back in Florida. Armstrong himself denies making any mistake: it is essentially deceived wisdom, he asserts – there was no blooper. At least one scientific analysis of the recording by an Australian computer programmer, Peter Shann Ford, in 2006 apparently supports Armstrong's insistence. Ford analysed and cleaned up the NASA recording of Armstrong's words using simple sound-editing software and says that the 'a' is very short, but very there. Other defendants of Armstrong's words blame his Ohio accent for the clipped 'a'.

However, a more recent analysis of the recording by Apollo expert Chris Riley and forensic linguist John Olsson has shown that there really is no 'a'. Armstrong did fluff his line, though the phrase is poetic nevertheless. It would have been a nerve-wracking moment, after all, and we can perhaps excuse the first man on the moon a little bit of stage fright.

But this little bit of deceived wisdom could prove to be the most important instance of anyone fluffing their lines ever. Why? Because it proves that Armstrong really did set foot on the moon. There are many people who believe the most bizarre and convoluted conspiracy theories. Among them are those unwilling to accept that anyone has taken even the smallest of steps, let alone a giant leap, on the moon. It is difficult to understand why.

Perhaps some of them saw the 1978 movie *Capricorn One* about a faked Mars landing and assumed it to be based on scientific fact rather than a work of fiction for the entertainment of cinemagoers. Others, it seems, take an Aristotelian stance of the kind dramatically overturned by Galileo and Kepler more than three centuries ago and brought to life recently in Stuart Clark's gripping novel *The Sky's Dark Labyrinth*. They assume that the solar system is somehow perfect, occupied by immutable spheres whose music resonates for all eternity rather than simply being orbiting chunks of gas and rock around the vast nuclear fusion reactor that is the sun.

The conspiracy theorists talk of 'evidence' to support the idea that the lunar landing was faked. They cite implausibly high photographic quality, problems with the position of camera cross hairs in the images sent back, the lack of stars in the lunar sky, inconsistencies in lunar shadows, photographic hot spots caused by alleged stage lighting and, of course, the Star-Spangled Banner purportedly rippling in the breeze. All this 'evidence' has been debunked repeatedly, but the theories won't go away.

It is Armstrong's fluffed line that might just do it. That missing 'a' is the single most critical piece of evidence we have with which to banish all the lunar conspiracy theories in a single syllable.

Picture the scene. It's 1969, and the US Government and NASA have commissioned you to direct the most elaborate movie ever. One that might change the course of human history. One that would be the culmination of cold war posturing. One that requires a giant leap of faith. It has to convince everyone of its authenticity, from Times Square to Red Square by way of Berkeley Square. It has to sing the story more sweetly than a nightingale.

You have to get everything right: script, scenery, continuity. The whole world and future generations have to be convinced that it is authentic, or all is lost. The big moment comes, and the cameras are rolling for that crucial scene. Your leading man, the man of the lunar matinee, is just about to plant his foot in the grey dust scattered across the studio floor. His monologue is good to go, his 'To be, or not to be' moment is cued ... and ... 'Action!'

'That's one small step for man, one ... giant leap for mankind.'

At this point, what would you, as director, shout?

'Cut!' That one word would be quickly followed by a rather irritated 'Steady, now, Neil, daaahling, back up those steps would you, be a love? Ready to go again everyone? *Lunar Landing*, scene 1, take 2 ... and ... action!'

'That's one small step for a man, one giant leap for mankind.'

'Perfect! That's a wrap everybody, the moon's in the can. See you in the bar – the drinks are on me!'

Faked to perfection. Of course, say the conspiracy theorists, the people who faked the moon landing would have thought of that, wouldn't they? Isn't that why they left out that little 'a'? Isn't Armstrong's fluffed line just the perfect touch to add nervous authenticity to the movie?

Next time you're gazing up at the night sky and musing on whether or not there is an astronaut's footprint on that great ball of lunar cheese, think on: if Armstrong had been word-perfect, we might never have been convinced of the truth.

• • • • • • • • • • • •

The Science

There is an abundance of evidence, including chunks of moon rock that NASA displays and shares with schools and museums, to prove that humans have been to the moon and come back down to earth. A scientific analysis of the most famous words not spoken on earth also proves that Neil Armstrong planted his feet and his flag on the surface of the moon.

* * * * * * * * * * * *

Find out more:

http://news.bbc.co.uk/1/hi/8081817.stm

Food fads and fertility

The Deceived Wisdom
Specific foods such as sardines can boost
a man's chances of becoming a father.

• • • • • • • • • • • •

As a regular science blogger I see a lot of odd search-engine hits on my website. People arrive looking for a wide range of things – from the obvious 'scientific discoveries' and 'free science magazines' to information about chemical compounds and answers to science homework questions. Others are more peculiar: 'spot the fake smile', 'five-leaf clover' and 'seven deadly sins', for instance.

The oddest of phrases recently brought a clutch of new visitors to the site, all looking for 'sardines and fertility'. Sadly my site didn't include an article on that rather specific subject, though there were news items in the archives that discussed contamination of fish samples by poisonous metals such as arsenic, lead and mercury. I wondered whether there might be some advocacy group touting sardines as a bizarre alternative to more technologically sound approaches to fertility such as avoiding tight trousers and hot baths, and getting enough vitamins and minerals. Perhaps the idea was started by the fishing industry to boost sales.

A detailed search of the medical research databases for scientific papers about fertility, sperm and reproduction in the context of sardines (and even pilchards) failed to bring up anything other than one journal article about the reproductive habits of the Spanish sardine, *Sardinella aurita*, to the southeast of Margarita Island, Venezuela. This was probably not entirely relevant.

There is some evidence that trace nutrients found in some foods can improve fertility. For instance, the vitamin-like nutrient coenzyme Q10 occurs in beef, soy, spinach, peanuts, vegetable oil and, yes, sardines. This compound is found mainly in the powerhouses of our body's cells, the mitochondria, and helps to release the energy from ingested food. One research study suggests that it might protect sperm from damage by oxidizing chemicals and somehow even help maintain sperm count or ensure that these gene-bearing cells are swimmingly mobile.

There is a lot of spurious information about coenzyme Q10 on websites that discuss ways to improve fertility naturally. Many of these do indeed mention sardines, but only in the context of long lists of foods that supposedly help fertility. Other sites simply suggest adopting a healthy diet and lifestyle to boost the chances of fatherhood. There are claims that the healthy fats, polyunsaturated omega-3 fatty acids, found in sardines are 'critical to fertility'. These fats do seem to be more generally beneficial to health in terms of reducing inflammation and staving off diabetes, but there is no strong evidence of their sperm-boosting abilities. Although at least one site claims that omega-3 fatty acids might boost libido.

One thing that is seriously overlooked in all the discussions of this piece of deceived wisdom is whether it is best to choose the sardines tinned in oil or in tomato sauce.

. . . .

The Science
While there is evidence that a balanced diet and healthy lifestyle may be beneficial to male fertility, there is no evidence that specific foods will solve fertility problems.

. . . .

Find out more

http://www.sharecare.com/question/
are-there-foods-that-increase-sperm-count

A miscellany of misconceptions

• • • • • • • • • • • •

CHLORINE MAKES SWIMMING POOLS SMELLY AND STINGS YOUR EYES – Various chemicals are regularly added to the water in swimming pools to ensure that it stays safe for swimmers. Some of these chemicals prevent algae from growing and turning the pool water green, while others kill the harmful microbes that find the warm water and warmer bodies of swimmers a wonderful place to make their microscopic homes. Among the most commonly used chemicals added to pools are solutions of sodium hypochlorite. This is the same active ingredient that is found in household bleach, and is also added at very dilute levels to the domestic water supply to keep it safe.

Sodium hypochlorite kills microbes by breaking down, or oxidizing, the molecules from which the microbes are composed. It is akin to a slow chemical burn and, as the advertisers say, it kills 99 per cent of germs. Of course, there are other molecules present in swimming pools, especially when swimmers are present. They include the oils and other secretions present in sweat and, most notoriously, 'urea' from the less hygienically aware swimmers. (Sadly, there is no dye that pool owners can add to pools to reveal embarrassingly who the urinating culprit is.)

Nevertheless, hypochlorite also oxidizes these compounds, producing a new group of chemicals that are more volatile, known as chloramines. It is the chloramines that get into the air around a swimming pool and give it that familiar 'chlorine' smell. It is these same compounds present in the water that sting many swimmers' eyes and are even thought to be responsible for some cases of asthma. The rules of the pool are not dreamed up just to annoy swimmers: the directives 'please take a shower before you swim' and 'no urinating in the pool' are in place for a very good reason.

• • • • • • • • • • • •

Find out more

http://www.cdc.gov/healthywater/swimming/pools/
irritants-indoor-pool-air-quality.html

SLAPPING SOMEONE ROUND THE HEAD DESTROYS 10,000 BRAIN CELLS – Many a youngster has attempted to persuade the school bully not to inflict a blow to their cranium with cries that they will lose 10,000 brain cells. First, this feeble excuse is unlikely to dissuade a bully intent on causing harm. Second, it is a classic misconception. The brain is well protected, by skin, hair (for those people more hirsute than this author) and a well-padded layer of bone, from all but the most severe blows or impacts targeted at weak points such as the temples. And because we do not know precisely how many brain cells we have – the number is estimated to be between

10 and 35 billion – the question of exactly how many might be lost following a slap to the head is moot. Who counted them before the slap and after?

IF ASTRONAUTS RIP THEIR SPACESUITS WHILE IN SPACE, THEIR BLOOD WILL BOIL – Many a sci-fi movie will see an astronaut accidentally or deliberately cut free from the spaceship, their spacesuit ripped in the process and the terrifying screams that no one can hear in space as their blood boils. Well, I'm sorry to say it's nonsense. This particular misconception presumably came about because, as every junior physicist knows, when you lower the pressure acting on the surface of a liquid it can boil at a lower temperature. It is one of the reasons why it is difficult to make a decent cup of tea at the top of Mount Everest: the low pressure of the atmosphere at that height reduces the boiling point of water to about 63°C, and the tasty and aromatic components of the tea leaves do not readily infuse into the water.

Of course, in the vacuum of space there is zero pressure, so one might imagine that liquid blood will boil. Fortunately for our clumsy astronaut, the pressure of blood itself, due to the pumping of the heart, is perfectly adequate to keep it liquid and prevent it from boiling. On the other hand, if the astronaut opens their mouth, then the saliva might boil away, though it would do so at body temperature and wouldn't scald the astronaut's mouth.

. . . .

Find out more

http://imagine.gsfc.nasa.gov/docs/ask_astro/answers/970603.html

THERE ARE WAYS TO TELL AN UNBORN BABY'S SEX WITHOUT ULTRASOUND OR GENETIC TESTING – There is a common misconception that you can tell the sex of an unborn baby from the direction of rotation of a needle, wedding ring or artefact, suspended by a thread above the expectant mother's outstretched palm, exposed stomach or some other body part. This is even less plausible if the woman is not yet pregnant. You cannot determine the baby's gender from the mother's stomach, how she stands, what cravings she has or even something as scientific-sounding as the baby's heart rate.

An ultrasound scan might give you a good view, but at the early stages of pregnancy the image can be ambiguous, and can remain so even at the late stages. A genetic sex test would prove it one way or the other (or somewhere in between), but is not recommended for this purpose, given the size of the needle required to access the baby's DNA.

· · · · · · · · · · · · ·

Find out more

http://www.babycentre.co.uk/pregnancy/naming/
knowing-gender-folklore/

It's brainy cats and dogs

The Deceived Wisdom
'Cats are far more intelligent than dogs' (the cat-lover).
'Dogs are far more intelligent than cats' (the dog-lover).

. . . .

The dog's amazing ability to behave in a social and sociable manner, the tricks new and old, the understanding of human language, the compassion and the empathy, and the whole 'man's best friend' thing are all testament to the dog's superiority over the cat in the brain department.

All domestic dogs, whatever their size and shape, are descended from an ancient canine ancestor closely related, if not identical, to the modern grey wolf. The domestic dog's scientific name gives us a clue: *Canis lupus familiaris. Canis* is Latin for dog (it is also the origin of the name of the Dog Islands, known to us in the English-speaking world as the Canary Islands). *Lupus* means wolf. *Familiaris* can mean so many things: familiar, domestic, servant, family. So our dogs are essentially family wolves.

The statuesque German shepherd, still occasionally referred to as an Alsatian? A descendant of that primeval wolf, and it looks like it. The golden retriever? Certainly. But what about

> **"**
> # I have studied many philosophers and many cats. The wisdom of cats is infinitely superior
> **"**
>
> Hippolyte Taine

the diminutive charms of the chihuahua, or the peculiar features of the pug? Both attributes of these wolves in deep disguise. And the St Bernard – which incidentally never did carry brandy in a miniature barrel around its neck on mountain rescue missions – that too? Sure. All domestic dogs today are descended from that ancient wolf species.

About 15,000 years ago, humans recognized a certain kinship with wolves and their pack-forming tendencies. Humans had already domesticated wolves, but none of those early breeds survived the last ice age. At the time, humans had no cities or farms, and were largely nomadic and constantly on the lookout for an easy meal – and for rival tribes who might try to steal their easy meal, or even their mates or offspring. Those humans from before the dawn of civilization saw that the grey wolves could work as a team, with a leader and subordinates and underlings, just as humans so often do. It was the classic alpha-male set-up. All members are afforded the protection of the pack and everyone gets fed, albeit some with fillet steak and others with gristly scraps and bones. Importantly, it's not just the alphas who get to spread their genes through sexual reproduction: the less dominant members, if they are wily enough, will find a way too.

Over the centuries we have bred dogs as hunters, retrievers, guards and companions. The most familiar modern breeds are all younger than a few hundred years old, though some breeds can be traced as far back as the ancient Egyptians and perhaps even earlier. But while they may look very different from one another, they are all still *Canis lupus familiaris*. They all generally share useful traits such as a powerful sense of smell, which generations of humans have put to use in hunting and searching. They can all be trained to respond to specific human commands, whether a sound, a word or a visual signal. In fact, there is evidence that at least some border collies, which are considered to be among the most intelligent of dogs and commonly accompany shepherds, can recognize and respond to several hundred different words. This is an intelligent beast.

Indeed, dogs can not only respond to human commands like any good servant, but they can also be highly mischievous and ignore those commands. Notorious for this is the Labrador retriever, which will, given a weak command, totally ignore it and take the opportunity to eat … something. Ultimately, though, it will succumb and bend to the dog handler's will. It will even show signs of remorse, knowing that it has disobeyed: ears down, tail tucked between its legs, head bowed. This suggests to some observers that dogs may have a strong sense of empathy – they know what you are thinking. The dog realizes it has done wrong and wants to make amends. Have you ever heard of a cat showing remorse?

But is remorsefulness an indicator of intelligence? The perceived wary aloofness of cats and their apparent ability to manipulate their owners to get what they want, whether that's food, an opened door or a fluffy ball of wool with which to play, often suggests to owners that cats have a far greater intelligence than dogs. A dog will repeatedly run after a thrown stick or ball, return it to the owner and allow the process to be repeated endlessly. One might wonder who gains from such a tedious pursuit. In the meantime, the cat will have found a comfortable spot in which to curl up and take a nice nap.

All domestic cats, whatever their size and shape, are descended from an ancient feline ancestor – *Felis silvestris lybica*. The domestic cat's scientific name gives us a clue: *Felis silvestris catus*. *Felis* is Latin for cat, and the origin of the name of a well-known cartoon cat from the mid-twentieth century, Felix. *Silvestris* means 'of the forest', and is also the origin of the name of a cartoon cat, Sylvester. And *catus* means, not 'cat', as you might at first think, but 'intelligent, clear-thinking and wise'. So, our cats are essentially wise cats from the forest. Sounds about right.

Cats have been with us for about 10,000 years in domestic form, and while the ancient Egyptians worshipped them as gods, for most of their time with us they have been used as living rodent traps, keeping the mice and rats away from our food stores and ankles.

The truth of the matter is not so much that either cats or dog are more intelligent than the other, but that each species is being successful in its own way. The ancestors of both cats and dogs evolved to exploit the environment in which they found themselves to obtain the food, shelter and mate they needed to successfully raise offspring and pass on their genes to the next generation. Humans are a late arrival on the world stage, and we often consider ourselves to have superior intelligence. But who is the more intelligent – humans, or the species that exploit us for free board and lodging, protection from rivals and predators, a never-ending supply of food and treats, the opportunity to play, and the chance to encounter members of the opposite sex in order to make new kittens or puppies so that the next generation can exploit us all over again?

If you really want to know, Susanne Shultz of the Institute of Cognitive and Evolutionary Anthropology at the University of Oxford has settled the age-old dispute about the relative intelligence of cats and dogs once and for all. Relative to body size, the dog's brain is much bigger than the cat's; moreover, the cat's brain is smaller than a cow's. Now, brain size does not directly reflect intelligence, but bigger brains usually do correlate with a greater ability to learn, come up with new behaviour, adapt to new environments, and so on. Cats may seem aloof, but that's simply because they're a bit dim. Even the dumbest dog can learn to chase a stick.

Let's not forget, however, that there are plenty of people around who do not consider dogs or cats either intelligent or particularly worthy of sharing their homes as pets. Those people know only too well that their pet snake/lizard/ferret/pig/stick insect (delete as applicable) is far more intelligent than any dumb cat or dog. At least their pet doesn't have to be kept entertained by humans doing tricks with dangling balls of wool or throwing sticks.

· · · · · · · · · · · ·

The Science

Despite their apparent aloofness and intellect, cats have a much smaller and more primitive brain than dogs and are fundamentally less intelligent. But cats get all the food and shelter they need without having to perform tricks or take on the social 'skills' of their canine counterparts. In that sense the two species are each perfectly adapted to their environment, both intellectually and physically.

· · · · · · · · · · · ·

Find out more

http://www.psychologytoday.com/blog/canine-corner/
201012/are-dogs-more-intelligent-cats

Infernal combustion
and the mobile phone

The Deceived Wisdom
You must switch off your mobile phone to prevent fires
or explosions at petrol stations should it ring.

. . . .

For the best part of two decades, well-intentioned drivers
with a strong social conscience have diligently adhered to
the golden rule of the garage forecourt: thou shalt switch
off thy mobile communication device before filling the tank of
thy vehicle. Of course, the less diligent boy (and girl) racers,
the wide boys and wise gals, and the flippant drivers of white
vans simply flip a digit at the sign and carry on their fascinating
phone conversations while filling up.

So, is there any risk of a stray radio wave from a mobile phone
blowing the user sky high if the diffusion of petroleum vapour
happens to coincide with that particular cellular airspace? Of
course not. It's just another piece of deceived wisdom. It has
never happened, and it's never going to happen. Never in the
twenty-year-plus history of the mobile phone has a filling station
burst into flames because someone answered their phone;

indeed, there has never been even the tiniest of conflagrations, despite what you may have read in hoax emails that have been circulating since the 1990s.

There are countless opportunities for sparks to ignite flammable vapours and liquids at any petrol pump, but you'll never see a sign warning you not to switch your vehicle's radio on or off – nor, for that matter, to start or stop your engine. Mobile phone masts might tower over a filling station where you decide to take that call from your mum, right at the moment the other drivers plunge nozzles into tanks, but there will be no fireball, no flash of white heat.

Vaporized petrol and diesel do burn, of course, otherwise all those internal combustion engines would be just so many inert blocks, crankshafts and pistons. However, it takes a very special combination of physical conditions to ignite the vapour. Diesel has to be hot enough, which is achieved by compressing the vapour and giving it a boost with the engine's glow plugs. A spark from a spark plug ignites petrol, but this is effective only if the vapour is contained in a small space. And whereas the 12 volts and large current (20 to 40 amps) from a car battery will achieve ignition, the 3 to 4 volts and tiny current (less than 1 amp) from a mobile phone battery will not.

Sparks come in all shapes and forms – and it's true that you can build up static electricity in the synthetic fibres of your clothes and from the plastics and fabrics of the interior of your car. However, you would have to be very deliberate in your actions to avoid this static simply being discharged to earth as you got out of your car vehicle and grabbed the fuel nozzle. And even if it were not discharged before then, it would certainly be grounded once you touched your vehicle's fuel cap before starting to fill up.

All emails, anecdotes and spurious media reports about mobile phones causing fires on garage forecourts are nothing more than deceived wisdom.

. . . .

The Science

Mobile phones do not cause fires at filling stations. There is no recorded fire department report of any such incident ever having happened anywhere in the world. There have never been any serious news reports claiming such a happening. The concentration of fuel vapours and the energy requirements for ignition are never met – except inside engines.

. . . .

Find out more

http://news.bbc.co.uk/1/hi/england/kent/4366337.stm

Physics falsehoods

• • • • • • • • • • • •

THE SPEED OF LIGHT IS CONSTANT – In the vacuum of space, light travels unimpeded at a very high speed, covering a distance of 299,792,458 metres in just one second. In fact, this is the universe's speed limit. According to Einstein, nothing can go faster than light in a vacuum, and the speed of light in a vacuum is constant for all observers. But light does not always travel at full speed. When it passes through a material, whether glass, water, air or any other transparent substance, its speed is lower.

Light slows down whenever it enters a medium other than a perfect vacuum. The rate of slowing is related to the refractive index of the medium. (The refractive index of a medium is defined as the ratio of the speed of light in vacuum to its speed in the particular medium.) For instance, the refractive index of glass is 1.5, so light travels through it at about two-thirds its vacuum speed, at about 200,000,000 metres per second. The refractive index of air is very close to 1, just a fraction over at 1.0003. Air therefore slows light only very slightly, by 90,000 metres per second, to about 299,700,000 metres per second.

There are materials that can slow light to almost human speeds. Chill a gaseous cloud of rubidium atoms to close to absolute zero to produce a form of matter – a so-called Bose–Einstein condensate (BEC) – and you have a substance that is very different from the familiar gases, liquids and solids we know at normal temperatures. A BEC can slow light to just a few metres per second, which is well within the speed limit of many roads. But, of course, you couldn't drive your car through a BEC, and it would be rather chilly at close to absolute zero!

* * * * * * * * * * * * *

Find out more

http://math.ucr.edu/home/baez/physics/Relativity/
SpeedOfLight/speed_of_light.html

THE LAWS OF FLIGHT PROVE THAT BUMBLEBEES CAN'T FLY – Bumblebees, of course, can fly very well, their yellow and black stripy and rotund bodies manoeuvring easily from flower to flower. The high-frequency flapping of their two pairs of wings gives rise to their familiar buzziness. Aerodynamic calculations that assume a bumblebee to be a rigid object that is relatively heavy in relation to its stubby and stiff wings would certainly show that such an object cannot fly. That 'model' of a bumblebee may be grounded, but no aeronautical engineer has ever suggested that a real bumblebee cannot fly. The very latest mathematical models of the little buzzers show that they have flexible wings, which work in concert to give the bumblebee its high-speed aerodynamic grace by slicing through the air at a high angle and creating a vortex at their leading edge to help keep the bees airborne.

• • • • • • • • • • • •

Find out more

http://www.sciencenews.org/view/generic/id/5400/title/
Math_Trek__Flight_of_the_Bumblebee

YOU CAN TRAVEL INTO OTHER DIMENSIONS THROUGH A BLACK HOLE – Einstein's theory of relativity helps to explain the structure of space and time, the speed of light, and the gravity that keeps our feet on the ground and our planet in its orbit around the sun. It also suggested the possibility of wormholes – short cuts that might form between two distant regions in space-time. Calculations using theoretical physics once suggested that a certain type of infinite black hole could act as a wormhole, and science fiction authors have exploited the idea endlessly as time-travel portals. Unfortunately, entering a black hole would be a rather unpleasant experience to say the least, with crushing gravitational forces tearing you apart and incredibly intense X-rays frying you and your spaceship. It would also be entirely in vain, as the old calculations have now been overturned by modern physics, which shows that black holes cannot create wormholes.

• • • • • • • • • • • •

Find out more

http://helios.gsfc.nasa.gov/qa_sp_bh.html

Addicted to addiction

The Deceived Wisdom
Bad habits and compulsions are
the same as addiction.

. . . .

The tabloid media seems to be hooked on addiction. In 2011, singer Amy Winehouse was found dead in her home at the tender age of 27 years and so tragically joined many other musicians, including Jimi Hendrix, Kurt Cobain, Jim Morrison and Janis Joplin, as the latest recruit to the 'Forever 27 Club', a name that glamorizes their deaths and apparently bumps up their CD and download sales. Not long after, the tabloids became preoccupied with the death of a young man who suffered the fatal consequences of deep-vein thrombosis, having spent more time than was really good for him playing video games day in, day out.

It's easy for the columnists to hook a rant on the notion of addiction. After all, there are so many things to be addicted to, aren't there – heroin, cocaine, tobacco, alcohol, painkillers, caffeine, chocolate, potato chips, aerobics, zumba, bungee jumping, golf, sex, work, pornography, even the internet and

Twitter or Facebook. The media, and consequently people in general, will talk flippantly of workaholics, sex addicts, chocoholics and other types of addict.

Technically, however, many of these urges so glibly labelled as addictions are in fact just bad habits to which those who yield feel they have surrendered, and not truly addictions in the sense of brain chemistry. While it may be tempting to say that someone who eats a large amount of chocolate or drinks a lot of coffee is somehow an addict, addiction is much more complicated than that.

The notion that simply using a harmful drug – cocaine, alcohol or whatever – leads to an inevitable descent into addiction is another piece of deceived wisdom. Many people 'use' without ever becoming addicted to their 'drug' of choice, whether that's a spin on the overtime hamster wheel at work, the exercise bike or the roulette wheel. They might even reach the point where their habits consume more of their time, money and concentration than any other activity in which they take part. But even then, taking their 'drug' to excess does not necessarily make them addicted in the clinical sense.

So, what is addiction? Unfortunately it is almost impossible for medical scientists to agree on a definition. Not everyone who indulges in a drug habitually is addicted, but all addicts use their drug habitually. Is addiction just tolerance, where to continue to experience the same 'high' the person needs a gradually increasing level of exposure to their drug of choice? Or is it dependence, where absence of their drug leads to the unpleasant symptoms of withdrawal? It is difficult to know whether there is actually a difference between what one might call psychological addiction and physiological addiction.

Many users keep on indulging their habit to avoid the

unpleasantness of not taking the drug rather than the pleasure of taking it. That seems to apply whether they are smoking cigarettes, snorting cocaine or shuffling virtual chips in an online poker game. It is impossible to state with clarity whether indulgence is addiction – does the depth of the 'lows' between the 'highs' influence the definition?

Reaching a position in which an uncontrollable chemical imbalance in the brain continually drives a person to seek their fix is something from which no one is immune. We all have 'reward' centres in our brains that can respond to stimuli from 'reward' brain chemicals like dopamine, which are released to give us pleasure. The stimulus–reward feedback is reinforced to different degrees by different stimuli. Drugs of abuse that intervene or modify these reward pathways in the brain can modify brain chemistry directly.

However, we all have our bad habits and irrepressible urges. But the flippancy with which terms such as 'workaholic' and 'sex addict' are bandied about in the tabloid media belies a much more insidious problem: the disease we refer to as addiction destroys lives in ways that are way beyond such flippancy.

Addiction does seem to go hand in hand with certain mood and anxiety disorders. It is not clear whether one leads to the other or whether there is a direct causality. Richard A. Friedman writing on Amy Winehouse and addiction in the *New York Times* in August 2011 concluded that genes, environment and psychology are all at play in determining who will become addicted to what.

Perhaps that is obvious. There is no way to unravel the complex relationship between one's genes and one's circumstances and environment. And it makes no sense to talk of the psychology of addiction without discussing its neurochemistry and the way in which the reward pathways might be programmed by our genes or even environmental effects early in life. What is clear is that the tabloid hyperbole and scaremongering surrounding those illicit drugs that society fears the most needs a detailed reassessment, as does the nature of addiction in areas such as gambling and other non-chemical stimulants. This is especially important given the flippancy with which the terminology of addiction is applied to problems that seem not to be related to addiction but are simply bad habits.

* * * * * * * * * * * * *

The Science
Despite the tabloid media's claims about sex, drugs, rock and roll, and even chocolate, having a bad habit or craving is not necessarily an addiction.

* * * * * * * * * * * * *

Find out more

http://www.psychologytoday.com/blog/
inside-the-criminal-mind/201102/
the-overuse-and-misuse-the-word-addiction

"

If the facts don't fit the theory, change the facts

"

Albert Einstein

The three Rs of sustainability: Reduce, Reuse, Recycle

The Deceived Wisdom
Recycling is a waste of time and energy.
It's simpler, cheaper and more efficient
to dump waste in landfill.

. . . .

ENERGY – Those opposed to recycling often say that it takes more energy and resources and produces more of the greenhouse gas carbon dioxide to collect, process and recycle glass, paper, metal and plastic. This is dangerously misleading deceived wisdom. Mining metal ores, running oil wells and converting crude oil into plastics, felling trees and pulping wood all use far more energy than the relatively simple collection and processing of waste products.

GLASS – Only a fraction of glass bottles can be recycled into new containers. Colourless glass is suitable, but unfortunately brown and green bottles are less easy to recycle, and produce low-quality recycled glass. Nevertheless, waste glass can still be recycled in other ways, by grinding it down to make industrial abrasives, adding it to building and road construction hardcore and aggregates, or using it as filler for concrete and ceramics. This might sound as though the glass is being wasted, but the production of virgin glass is an energy-intensive process, as is the production of cement, and both also release vast quantities of carbon dioxide. By using waste glass aggregates and concrete, the need for fresh raw materials for those applications can be reduced. So all glass is recyclable, just not into new bottles.

PLASTICS – Some modern recycling centres have highly sophisticated sorting systems that use laser-guided conveyors and air-jet systems to separate different plastics in a stream of shredded mixed waste. Laser beams scan the plastics and a spectrometer reads their chemical signature, much as a laser scanner at a checkout scans the barcodes on the items from your shopping trolley. A puff of air is then sent from the appropriate jet to send each shard of plastic into the appropriate hopper. As this technology gets cheaper and more widespread, the ability to extract different plastics from household and industrial waste will become greater, and less plastic will be sent to landfill or wasted.

ALUMINIUM – Aluminium was once considered to be a precious metal, as it was so expensive to extract from its ore, bauxite. The statue of Eros in Piccadilly Circus, London, is made of aluminium, as is the cap on the top of the needle-shaped Washington Monument in D.C., such was the esteem in which this metal was once held. Indeed, the cost of making aluminium during the nineteenth century made it more expensive than gold. The cost fell when the electricity needed to extract aluminium from bauxite became cheaper to generate. Today, lightweight aircraft and supercars aside, we tend to use the metal for more mundane purposes, such as cans for drinks and 'tinfoil' sheets for wrapping the Sunday roast.

Electricity is still relatively cheap (despite recent price rises), but the energy costs and environmental impact of strip-mining bauxite and extracting the metal are very high. Melting aluminium requires less than 400 kilojoules of energy per kilogram of metal. Breaking down aluminium oxide (bauxite) uses about eighty times as much energy. So collecting and melting aluminium cans, and purifying to recycle the metal, uses, overall, about 5 per cent of the energy needed to produce virgin metal from bauxite.

PAPER – Paper is made by pressing wet, pulped fibres of cellulose from wood, sugar cane or grasses into sheets and then drying. Should this book ever stop selling, the remaining handful of copies in the publisher's warehouse may well be pulped (perish the thought) and added to the stockpile of recycled paper ready for conversion into the daily newspapers, which millions of people do still read despite the advent of electronic news and the web. Only about one-third of the trees harvested each year are used to make paper; the majority are used in manufacturing wooden products, in construction and other areas.

ORGANIC MATTER – Until recently, organic waste was simply sent to landfill, and the methane gas released as the waste rotted underground was vented or burned off. But many recycling and composting centres can now accept all kitchen and garden waste and process it into useful compost. The advent of enclosed sorting and composting centres allows vermin-attracting kitchen waste to be dealt with, whereas earlier systems could cope with garden waste only. Some facilities now even 'compost' mixed refuse to extract the organic matter as a liquid run-off.

LANDFILL – Today, sending waste to landfill should be the last resort, once plastics, metals, paper and organic waste have been extracted. There are now numerous facilities in place to collect textiles and other recyclable materials or for them to be sorted on site. Yet even then, that's not necessarily the end of the matter. It is deceived wisdom that once buried in landfill, waste materials are gone for ever. This is not the case now, and in the future there will be more and more effective ways to 'mine' landfill sites, digging up and processing old waste to extract precious metal, plastics and other materials which can then be recycled into new products.

. . . .

The Science

We may not be reducing our consumption, but there are many effective ways to reuse products and materials, and when they have finally outlived their usefulness they can be recycled. The three Rs are almost always the best option – certainly better than sending the waste to landfill.

. . . .

Find out more

http://www.popularmechanics.com/science/
environment/recycling/4290631

Theory under pressure skates on thin ice

The Deceived Wisdom
Skaters can skate on ice because their skates melt the surface of the ice, creating a thin lubricating layer.

• • • • • • • • • • • •

Countless science teachers, textbooks and fans of ice skating will tell you that ice melts under pressure. They explain how applying pressure lowers the freezing point of water so that it has to be much colder before it will freeze into solid ice, and how, conversely, ice under pressure will melt. The classic example of this phenomenon in action can be seen every time a skater's blades swish across the surface of an ice rink. The relatively sharp edge of the blade and the weight of the skater pressing down on the ice lower its freezing point so that the ice beneath melts, forming a thin film of liquid water on the surface of the rink across which the skate can then glide with almost no friction.

Unfortunately, in common with the deceived wisdom that adding salt to the roads in freezing weather will prevent ice from forming, it is simply not true. Adding salt lowers the freezing point of water slightly, but if the road is cold enough ice will still form.

The behaviour of water has puzzled scientists for centuries. For instance, unlike almost any other material it expands when it freezes. This is why titanic icebergs float: ice is marginally less dense than water, though most of an iceberg will remain below the waterline because the difference really is small. It also means that fish can still swim in lakes and rivers that appear to be frozen – because bodies of water effectively freeze from the top down, allowing water to remain liquid below the icy surface.

Water is also unusual in that we can experience all three common states of matter – solid, liquid and gas – within an accessible temperature range, from 0°C to just 100°C. By contrast, the oxygen in the air we breathe is commonly a gas, becomes a liquid only at about 183 degrees below zero and freezes at an even chillier minus 219°C. At the other extreme, common table salt, sodium chloride, is a solid until it reaches more than 800°C and vaporizes only at a scorching 1,413°C.

But back to those ice skaters. Scientists have calculated the change in the freezing point of water at different pressures and backed it up with experiments. To lower the freezing point of water from 0°C to –1°C you must apply a pressure more than 121 times the pressure of the atmosphere bearing down on your head right now.

One of the scientists who has done the watery sums, chemist Kevin Lehmann of the University of Virginia, has a solid answer. He started by assuming that an ice skater weighs about 75 kilograms. The blade of each skate in contact with the surface is about 3 millimetres wide and 200 millimetres long. Pressure is defined as the force applied to a specific area. The force pressing down on the ice is 75 kilograms multiplied by gravity, which has a value of 9.8 newtons per kilogram. (The newton is the unit of force, named after Sir Isaac Newton, the original spin doctor of falling apple fame.)

So, a force of 735 newtons (75 multiplied by 9.8) is pressing down on the blade of that skate. To calculate the pressure we need to know the area of the blade in metres. 3 by 200 millimetres is 0.003 by 0.2 metres, an area of 0.0006 square metres. The pressure is the force, 735 newtons, divided by this area, which is 1,225,000 newtons per metre squared, or 1,225,000 pascals. (The pascal, the unit of pressure, is named after the French scientist Blaise Pascal, who invented a mechanical calculator in 1642, when he was 19 years old.)

That value for the pressure being applied through the skate to the ice sounds enormous. And in some senses it is: it is about twelve times the value of atmospheric pressure, and it's the pressure you experience if you dive under water to a depth of about 120 metres. But this pressure is about ten times too small to melt ice. The skater would have to apply a pressure of 120 times atmospheric pressure to do that, and to exert that amount of pressure they would have to weigh ten times as much as a normal ice skater, and so be about 750 kilograms.

You might be wondering whether the pressure could be increased by sharpening the blades. After all, dividing the force of a 75-kilogram skater by a smaller number would equate to a higher pressure. But the effect would be to lower the freezing point of the ice only by a few tenths of a degree. Given that most ice rinks freeze their ice to well below 0°C, this would have little impact. The ice would stay solid.

So how do skaters skate over the solid and rough surface of ice if there is no liquid lubricant in the form of water to allow them to do so? Lehmann concedes that, as with many other properties of water, scientists simply don't know. There are theories about the water molecules at the surface and how they are not being held as tightly in the solid ice as those within the frozen solid. There are also ideas about defects in the structure of ice that might allow some water molecules to become loose and so enter the liquid state. It might be that the steel of the blade somehow grabs these loose water molecules and promotes melting as more and more water molecules loosen their grip on the ice to form that thin slippery layer of water below the skate. Either way, this has nothing to do with the pressure applied.

One property that quickly becomes apparent to anyone new to ice skating, however, is that when you land on the ice with a bump and struggle back to your feet, your body heat allows the frozen particles of ice to quickly revert to the liquid state … leaving you with a soggy behind.

· · · · · · · · · · · · ·

The Science

Applying pressure to ice has the effect of lowering its freezing point, which means it will melt to form liquid water above a certain temperature. However, the pressure exerted on the ice by even the bulkiest of skaters will be a fraction of that needed to melt ice at the frozen temperature of an ice rink.

· · · · · · · · · · · · ·

Find out more

http://www.faculty.virginia.edu/lehmannlab/badchemistry.html#ice

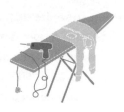

Multitasking men

The Deceived Wisdom
Women can multitask; men can't.

• • • • • • • • • • • •

For some reason there is a misconception that the male of the human species is unable to multitask, while the female has it down to a fine art. This is patently untrue of either sex. Drivers of both sexes, for instance, can cope perfectly well with keeping a vehicle on the road while simultaneously holding a conversation with a passenger and telling the kids in the back to be quiet.

There are countless other examples of how we multitask. Some tasks are inherently automatic, such as blinking, breathing and digesting your last meal; others are a little less vital, such as keeping an ear out for your significant other returning home, and (perish the thought) perhaps watching TV or browsing the web as you read this mighty tome. So we all multitask to some degree. Perhaps the misconception arises because many of us would rather run a shorter 'to do list' than juggling lots of different tasks.

In 2010, neuropsychologist Keith Laws and colleagues at

the University of Hertfordshire carried out tests on fifty male and fifty female students. They gave them all a couple of tasks to do, including trying to solve a simple puzzle while carrying out a map-reading exercise at the same time. Laws found no convincing evidence that anyone, male or female, could do three (or even two) or more complicated tasks at once, except for things involving autonomic responses, such as breathing, digestion and spontaneous reactions to stimuli, such as blinking when something moves rapidly towards one's face.

Even seemingly complicated tasks, such as driving, essentially become automatic after many months or years of practice, so most people can multitask when it comes to singing along to a tune on the radio while driving. But can you picture yourself solving a Sudoku puzzle, or following the plot of a crime drama, while driving and singing along? Probably not (at least I hope not, for the safety of our roads).

There are, it seems, tasks that we can do in parallel if those tasks are different enough. Singing while driving is one example. Driving is a learned procedural motor skill that stays with us – even in someone suffering the most drastic amnesia – while people with speech difficulties caused by brain injury or stroke can often still sing. It's perhaps no surprise that people can sing and drive at the same time, as neither task actually requires much conscious effort at all. Even watching TV and solving a crossword are perhaps different enough that we can 'multitask' such activities. But Laws's notion of high-level tasks involves being able to focus equally well and simultaneously on solving a Sudoku puzzle and playing a game of speed chess, for instance.

So is there any resolution in sight to this long-running battle of the sexes? One aspect of Laws's preliminary work on multitasking in men and women does lend some support to the notion that women are better at working on one task while planning another at the same time. Just over two-thirds of the women in the study performed better than the average man who took part in the tests. Laws found that some of the female participants could focus on thinking about how they would search for a lost key while capably answering a telephone call or working on a maths problem. He adds that there is now a small amount of evidence that a few female students can apparently work on two such tasks slightly better than their male counterparts. However, he points out that there is not one shred of evidence to confirm the converse: that men offset their lack of multitasking skills by being more focused. They are not.

Of course, the multitasking debate is not really about ability, but inclination – especially when the required multitasking involves housework, cooking, looking after children and holding down a full-time job, as opposed to sitting, drinking beer, eating pizza and TV channel-hopping – which is multitasking that many men and women are equally good at.

• • • • • • • • • • • •

The Science
Gender doesn't matter when it comes to combining simple tasks – or of course autonomic tasks such as breathing, blinking and digesting lunch. However, high-level multitasking in which two or more complex tasks are the focus is impossible for anyone, man or woman.

• • • • • • • • • • • •

Find out more
http://www.bbc.co.uk/news/magazine-11035055

Dozy
deceptions

• • • • • • • • • • • • •

YOUR BRAIN AND BODY SHUT DOWN WHEN YOU
SLEEP – You may think that when you drift off into
the land of nod, because you are not conscious your
brain and body systems have also dozed off. But this perception is as far from reality as you could get. Not only do you
keep breathing and your heart keeps beating, but you continue to digest food and produce urine. Some brain activity
actually increases during sleep as thoughts are processed
into memories, and you may even regain some awareness
in lucid dreaming during certain phases of sleep. However,
when you sleep your muscles enter a state of hypotonia (low
muscle tension). This seems to be a protective mechanism to
preclude potentially hazardous movements if you act out your
behaviour in a dream, for instance.

• • • • • • • • • • • • •

Find out more
http://www.nhlbi.nih.gov/health/public/sleep/yg_slp.htm#myths

63

IT IS DANGEROUS TO WAKE A SLEEPWALKER – Somnambulists can occasionally hurt themselves and others, so the reverse is true. Waking a sleepwalker could save them from embarrassing situations or, worse, falling down the stairs. It may confuse the person if they are woken and find themselves out of bed, but that will cause them no physical harm.

· · · · · · · · · · · · ·

Find out more

http://www.theregister.co.uk/2007/01/27/
the_odd_body_wake_sleepwalker/

YOU NEED LESS SLEEP AS YOU GET OLDER – Unfortunately, this is not the case: you need as much sleep as you need, and that does not get less as you get older. But what often happens is that you begin to suffer from insomnia or aches and pains that prevent you from falling into a deep sleep. Staying up late and rising early is a common way of avoiding a restless and uncomfortable night, rather than an indication that you no longer need the same amount of sleep.

· · · · · · · · · · · · ·

Find out more

http://www.eurekalert.org/pub_releases/2012-03/aaos-ssg022712.php

EATING CHEESE BEFORE BEDTIME GIVES YOU NIGHTMARES – It is often said that eating cheese before bedtime will lead to a fitful night's sleep full of bad dreams. The mean-spirited Scrooge from Charles Dickens' *A Christmas Carol* blames the appearance of ghosts on Christmas Eve on a

"

The best cure for insomnia is to get a lot of sleep

"

W. C. Fields

full meal including a 'crumb of cheese'. Back in 2005, the British Cheese Board attempted to overturn this particular crumb of deceived wisdom by giving 200 willing volunteers different cheeses to eat and asking them to report on whether they then had nightmares. The volunteers reported weird and colourful dreams, but no nightmares. The Cheese Board study even drew some spurious link between the type of dreams and the type of cheese, but there was no control group against whom to compare the number of dreams reported by people who didn't have the cheese.

Slightly more independent research suggests that the amino acid tryptophan, present in milk proteins in small quantities, actually improves sleep, though the amount present in cheese is far less than in meat. So cheese before bedtime, while not likely to stir nightmares, won't help you sleep any better than other foods – particularly if it gives you insomnia-inducing indigestion.

· · · · · · · · · · · · ·

Find out more

http://www.merseysideskeptics.org.uk/2011/09/
on-cheese-sleep-and-nightmares/

SNORING IS A NATURAL PART OF SLEEP – It probably will not help those who (try to) sleep in the same bedroom as a snorer to know that snoring, while common, is not really a normal part of sleep. It usually indicates some kind of obstruction of the airways, and that is rarely a good thing. What's more, loud snoring night after night is often associated with a condition known as sleep apnoea in which the person stops breathing for short periods of time. This condition can preclude a good night's sleep for both the snorer and anyone sharing the room, and there is also increasing evidence that sleep apnoea is a risk factor for high blood pressure and even type 2 diabetes.

• • • • • • • • • • • • •

Find out more

http://www.webmd.com/sleep-disorders/guide/sleep-fact-fiction

EVERYONE NEEDS EIGHT HOURS' SLEEP – This is just plain wrong: some people do get away with less than eight hours, though they may often be chronically tired without realizing it; others, especially teenagers, sleep more than that. One way to determine how much sleep you 'need' is to check how long you sleep when there is no alarm, work or pressing engagement to wake you. If you wake up after having slept for nine hours then you might assume that this is how much you need, though you could simply be compensating for a chronic lack of sleep over the course of several weeks. On holidays there are often exhausting sporting, walking and drinking activities to take into account.

We do not know how much sleep our prehistoric ancestors had, but evidence from recent history and from studies of remote tribes does suggest that natural sleep patterns tend to fit around the hours of daylight, so will obviously vary throughout the year. What we do know is that sleep deprivation can cause serious health issues, whatever you may hear about certain famous names, such as former British prime minister Margaret Thatcher and American inventor and businessman Thomas Edison both supposedly needing just three or four hours.

Some have suggested that Mrs Thatcher was not necessarily functioning at her best and might have been more effective in some of her duties if only she'd had more sleep.

There is recent evidence to suggest that, throughout most of history, humans have tended to sleep in more frequent and shorter bursts, spending their waking hours not worrying about insomnia. The eight-hour sleep shift seems to be a recent phenomenon.

· · · · · · · · · · · · ·

Find out more

http://psychcentral.com/lib/2011/
do-you-believe-these-10-sleep-myths/

IF YOU DIE IN A DREAM, YOU DIE IN YOUR SLEEP – How would anyone know what the recently departed were dreaming of before their demise?

· · · · · · · · · · · · ·

Find out more

http://healthynole.fsu.edu/Common-Health-Myths/Sleep-Myths

The most embarrassing sting

The Deceived Wisdom
Urinating on a jellyfish sting is the best
antidote to the creature's venom.

. . . .

Ah, seaside holidays … sunny climes and balmy waters, where the sand is golden and the sunlight glistens on the azure sea. Unfortunately, the warm waters lapping at sun-drenched shores are attractive not only to sun worshippers and junior sandcastle construction workers, but also to translucent creatures with trailing tentacles – otherwise known as jellyfish. The scientific name given to these creatures is *medusa*. As did their namesake from Greek mythology, these animals come with a nasty sting.

Jellyfish carry toxin-bearing sting cells that are triggered by close contact with the sensitive feet and legs of paddlers and swimmers attempting to cool off after a tiring afternoon building sandcastles or playing ball on the beach.

A jellyfish sting can be very painful, like a really nasty wasp sting, and in some cases much worse. Thankfully rare on holiday shorelines outside the tropics are species that have potentially lethal stings, such as the various species of box jellyfish, also known as sea wasps. Anyone who has visited Australia's Northern Territory will be familiar with the protective nets used to protect swimmers from these creatures.

There is one piece of advice that visitors to some tropical regions may have heard, perhaps in 'Strine' from a local lifeguard should they have strayed into the wake of a sea wasp (or they may have picked it up from a certain episode of the American sitcom *Friends*). The advice is that urine is the best antidote for a jellyfish sting because it purportedly contains chemicals that can neutralize the venom. Failing that, ammonia solution is occasionally recommended, though rarely offered.

There is obvious potential for embarrassment in using urine to treat jellyfish stings, and one has to question the motives of a virile young lifeguard who offers victims the chance to receive such personal medication. One must also consider the bacteria that urine might accumulate on its way from bladder to sting victim. There is a piece of deceived wisdom about urine which says that the liquid is sterile. It may well be while in the bladder, but once it leaves that receptacle this is no longer necessarily the case: there are plenty of bacteria that can line the tubes leading to the exit, so there is some risk of transferring problematic bacteria to the patient.

Regardless of the potential for embarrassment or the risk of infection, there is little point in using urine to treat a jellyfish sting. Fresh urine is usually neutral, chemically speaking – neither acidic nor alkaline. The waste liquid is about 95 per cent water and contains only a small amount of urea, a compound of nitrogen which is broken down by bacteria to release ammonia. This may be the origin of the claims that urine might be useful in treating a jellyfish sting, as in the past ammonia solution has been used to do so. Unfortunately, as with wasp stings, jellyfish stings are themselves alkaline, like ammonia solution. Vinegar, which contains acetic acid, is the safest liquid one might administer, but while it would be relatively easy to get hold of a bottle of it in a British seaside town, it is not usually quite so readily available on the tropical beaches where the jellyfish are more of a threat. However, serious cases should be dealt with by a doctor or professional healthcare worker as there is a risk for some people of a severe allergic reaction to the venom.

The deceived wisdom about jellyfish and urine is more than a problem of beachside embarrassment, though. The action of urine on the piercing sting cells that become embedded in the victim's skin is to trigger those cells, the so-called nematocysts, to release their venom into the skin. Similarly, applying surgical alcohol or even just rubbing the site of the sting can do the same. Rinsing with seawater would be a simpler remedy.

Fortunately, the stingers of some of the more hazardous and potentially lethal box jellyfish are very short. Lifeguards in those parts of the world where the waters are often infested with the stinging creatures therefore now have a much more straightforward solution to the jellyfish problem. Rather than suggesting urination as a treatment after being stung, they suggest a simple preventive measure that will protect the parts of your body most likely to come into contact with jellyfish tentacles. You should put on a pair of tights before going swimming: this will prevent those little stinging cells getting to your legs and bottom, at least. A full body stocking and swimming cap would offer even better protection – though whether that is any less embarrassing than having a lifeguard administer the traditional jellyfish treatment is a moot point. And if you choose fishnets, you risk suffering both embarrassment and jellyfish stings.

• • • • • • • • • • • •

The Science

Urine is usually a neutral or slightly alkaline liquid and so does not counteract the alkaline jellyfish sting. And urine can trigger the stinging cells to inject their venom more ferociously into the victim, making matters far worse. A much better solution is to douse the site of the sting with acidic vinegar or simply rinse with seawater. Never rub the skin.

• • • • • • • • • • • •

Find out more

http://www.scientificamerican.com/
article.cfm?id=fact-or-fiction-urinating

Seconds of snack-sized dietary deceptions

· · · · · · · · · · · ·

HOT, SPICY FOOD WILL GIVE YOU AN ULCER – The growing popularity of spicy food in Britain over the past few decades has given rise to an odd piece of dietary deception: that somehow hot spices cause gastric or duodenal ulcers. Fortunately for curry fans, this is simply not the case. The vast majority of duodenal ulcers (90 per cent) and gastric ulcers (60 per cent) are caused by a chronic infection by the bacterium *Helicobacter pylori*. This corkscrew-shaped microbe burrows into the lining of the gastrointestinal tract, causing inflammation and disrupting the acid balance and the natural protection of the lining, which leads to erosion and ulcer formation.

Scientists Barry Marshall and Robin Warren proved this to be the case and showed that a course of strong antibiotics could eradicate the infection and eliminate the risk of ulcers. Indeed, Marshall deliberately infected himself with *Helicobacter*, developed an ulcer and then cured himself with antibiotics.

Despite this very personal piece of research, the claims were met with scepticism as it was commonly thought that diet

– and stress – might be to blame for the development of ulcers. The first drugs to make billion-dollar profits were for treating ulcers by reducing the amount of digestive acid manufactured by the stomach. If simple antibiotics could be more effective at curing the problem, the pharmaceutical industry might lose some of its big money-spinners, as antibiotics are relatively cheap and generate little profit for manufacturers. Fortunately for those with ulcers, Marshall and Warren were vindicated and received the 2005 Nobel Prize in Physiology or Medicine for their research.

While not all ulcers are caused by *Helicobacter* (some are due to prolonged and excessive use of certain types of painkiller such as aspirin and ibuprofen), there is growing evidence that, rather than causing ulcers, some spices actually lower the risk of infection by *Helicobacter* in the first place.

· · · · · · · · · · · · ·

Find out more

http://www.nobelprize.org/nobel_prizes/medicine/laureates/
2005/press.html

SWALLOW CHEWING GUM, AND IT WILL GET STUCK IN YOUR GUT FOR SEVEN YEARS – Thankfully not. Gum will pass through the gut just as quickly as any other indigestible foodstuff, such as plant fibre. It does not take the seven years that this piece of dietary deception suggests. While chewing gum and bubblegum may seem like fairly modern inventions, humans have been chewing the rubbery, resinous latex sap from certain species of tree for thousands of years, and though modern gum often uses synthetic rubber it will still pass through you if you accidentally swallow it. There is a small risk of compacted chewing gum building up in the gastrointestinal tract, and that can cause pain and swelling; this is rare, but has been seen in children who have swallowed five or six pieces of chewing gum every day for many weeks or months. The result is the formation of a bezoar (a large indigestible mass in your gut), but the ill effects of such a mass developing would become a problem well before seven years were up.

• • • • • • • • • • • • •

Find out more

http://www.dukehealth.org/health_library/health_articles/
myth_or_fact_it_takes_seven_years_to_digest_chewing_gum

EATING CRUSTS MAKES YOUR HAIR CURLY – Parents often encourage their offspring to eat the crusts of their sandwiches with the incantation that doing so will encourage their hair to curl. The original idea was presumably to ensure that the children were getting adequate nutrition and plenty of 'roughage' (usually known as fibre these days), and hinges on the notion that curly hair is more desirable.

Of course, whether one's hair is curly depends almost entirely on genetics: your genes dictate the shape of your hair follicles. A perfectly round follicle will grow hair that is circular in cross section – such hairs will be less able to bend and so grow very straight. A more oval or irregular follicle will tend to form a hair with a flatter cross section which can more easily form waves, kinks or curls. People with straight hair can override their genetics by breaking into their hair using heat and curlers, or using perming solution to rearrange the protein molecules that make up the hair and allow them to be curled. Similarly, the curly-haired can straighten their hair using heat or chemicals. Of course, a perm (or permanent wave) is anything but permanent, as the hair will revert to its normal straightness as it grows.

· · · · · · · · · · · · ·

Find out more

http://www.theregister.co.uk/2006/08/04/
the_odd_body_crusts_curls/

IT IS NOT GOOD FOR ONE'S DIGESTION TO EAT FRUIT FOR DESSERT – Countless New Age and alternative health-care practitioners have followed the advice of the late alternative-medicine guru Herbert Shelton, telling their clients to eat fruit only on an empty stomach and certainly not after a meal because the fruit interferes with healthy digestion, ferments in the gut, and causes bloating and poor absorption of vitamins. Thankfully for fruit fans, the gastrointestinal tract is indifferent to the order in which food enters the stomach. It is best to take your fruit as you like it – which brings us to our final dietary deception for this chapter.

· · · · · · · · · · · · ·

Find out more

http://www.snopes.com/food/warnings/fruit.asp

AN APPLE A DAY KEEPS THE DOCTOR AWAY – This well-known saying suggests that eating an apple each day will keep you healthy and preclude the need for a visit from one's physician. Unfortunately, while apples are certainly a tasty and convenient fruity snack, they are very low in vitamin C and so cannot be safely relied upon to fulfil your daily requirements of that particular vitamin. A deficiency of vitamin C causes scurvy, which in French is *scorbut*, which has the same etymology as the chemical name for vitamin C, ascorbic acid. Of course, apples are good for your health, and as an extra bonus they contain other natural chemicals that help you to absorb vitamin C from other sources. Another plus may be that if you snack on an apple it deters you from eating sugary sweets instead. So perhaps an apple a day keeps the dentist away, if not the doctor.

· · · · · · · · · · · · ·

Find out more

http:// tlc.howstuffworks.com/family/
an-apple-a-day1.htm/printable

Charming
modern-day alchemy

• • • • • • • • • • • •

For centuries, the ancient alchemists worked mysteriously, in secret, brewing up strange concoctions in the vain hope of finding a substance that would bring the spark of life to inanimate matter. Mercury (or quicksilver), countless shimmering metals, magnetic lodestones, malodorous sulphur compounds, even more malodorous urine, eggs and human hair were boiled up in iron pots. The noxious and toxic vapours filled chambers lit only dimly by flickering candles or glowing furnaces.

It is a little bit of deceived wisdom that the alchemists' main aim was simply to get very rich, very quickly, by finding a way to turn lead into gold. Their true goal was to achieve immortality by creating a substance that would sustain life for ever, an elixir born of the dust from which, so the biblical texts told them, all life is made. As the scientific renaissance was brought to bear on the study of matter by Robert Boyle and his contemporaries, the influence of the alchemists faded and some dropped the 'al', which simply means 'the' in Arabic, to become ... chemists.

It is a modern tragedy that many vulnerable people, with a myriad of ailments and illnesses, are offered the modern equivalent of alchemical hocus pocus in the form of shiny amulets and magnetic charms. All too often, the sick and desperate resort to the mysterious in the hope that it will offer some relief. But Boyle and his successors have demonstrated that, although we may ultimately be no more than dust, the actual substances from which we are made – the proteins, the fats, the bone – are conglomerates of so many trillions of atoms – carbon, hydrogen, nitrogen, oxygen, and so on – all created in the nuclear furnaces of distant stars that exploded as supernovae and scattered their remains across the galaxy. We are stardust. That may be mysterious in its own way, but it is easily explained by science.

It is a sad fact that the purveyors of the modern-day philosopher's stone talk of imagined mysteries, such as auras surrounding people and life forces pumping through crystals. At best, they offer expensive but ineffective placebos to their customers; at worst, these talismans based on superstitions and pseudoscience distract the worried well and the truly sick from seeking attention for conditions that might be treated rationally with conventional medicine.

Today, sellers of crystals, copper bracelets, ionic wristbands, magnetic patches and the like make countless ephemeral claims for their products, though any actual health-giving effects are usually implied or simply hinted at. Indeed, as with much of complementary and alternative medicine, and as with the health food industry, and among makers of supposedly youth-giving cosmetics, it is in the practitioners' and manufacturers' interests to ensure that they make only superficial claims. This usually allows them to remain outside the jurisdiction of the medical regulators and stay on the right side of advertising laws.

The idea that living things have a vital force that exists beyond the physical world is an attractive one for many people – it is almost an instinctive feeling that harks back to our most primitive and primordial fears about life and death. In the nineteenth century, chemists such as the German Friedrich Wöhler synthesized urea and other 'natural' chemicals from existing chemicals in their laboratories. Until that time, scientists had assumed that chemicals from living things had some kind of distinguishing vital force which made them 'organic' – as opposed to 'inorganic' – so that they could not be made in a laboratory. Wöhler's laboratory experiments showed that there was nothing particularly special about the constituents of living things: they are just the same atoms and molecules, found throughout nature in the living and non-living world. In the years since, chemists have manufactured thousands of different molecules that were initially identified in living things.

Today, we can synthesize millions of organic molecules, including the material of our genetic code, the DNA. We can call those molecules that blend carbon, nitrogen, hydrogen and oxygen atoms 'organic', but that does not give them the vital spark that the alchemists sought.

- - - - - - - - - - - - -

The Science
Inanimate matter contains no life forces or
energies and cannot have an effect on health other

than by acting as a comforting placebo. However, such products often distract patients with serious conditions from seeking proper medical help that might successfully treat their symptoms.

. . . .

Find out more

http://en.wikipedia.org/wiki/Vitalism

Highly strung

The Deceived Wisdom
The classic Stradivarius violin has a unique sound
that justifies the reverence in which these instruments
are held and their million-pound price tags.

• • • • • • • • • • • •

Buying one of the 600 or so surviving violins made by Antonio Stradivari during the early eighteenth century will not make you sound better as a violinist. This is not because a better instrument will not improve your talents, but because there is no evidence that Stradivari's violins sound any different from those made by any other master craftsman. A Stradivarius violin is often associated with superlative excellence by players, composers and conductors. The sound, they say, is exquisite and cannot be reproduced even by the most beautifully crafted modern instrument.

There are numerous theories as to why a Stradivarius produces such a beautiful musical timbre. The design and shape of the violins and the craftsmanship used in their creation are considered the most important factors, but there are theories in support of others. In 2003, US scientists suggested that the wood available to Stradivari may have benefited from the Little Ice Age of the seventeenth century, making trees at the time grow more slowly than usual and so producing denser wood. Others have suggested that the sound might even be linked with the preservatives used at the time to kill woodworm and prevent moulds forming in the wood. There is even a theory that Stradivari used wood from ancient churches, endowing his violins with an almost spiritual quality.

Another focus of those hoping to explain the secret of the 'Strad' sound is that perhaps there is a secret ingredient in the varnish that somehow shapes the beautiful tone of these instruments. Researchers in Europe have smashed that notion, demonstrating that there really was no secret component in the varnish. They took microscopic samples of varnish from five Stradivarius violins and carried out a highly sophisticated chemical analysis. The instruments were made at different times over a thirty-year period. The analysis revealed that the varnishes contain materials, such as oils and pigments, that were used widely in decorative arts and paintings of the period. The researchers found no magic ingredient, no mineral or fossil resin layer, as had been suggested by some. There is, it seems, no secret sauce.

The Cremonese creator was certainly a master instrument-maker: he made hundreds of violins, violas, cellos, harps and guitars. The examples of these instruments that survive do indeed sound beautiful. Unfortunately for those clinging to the idea that there is something mysterious about these instruments, in this case beauty is in the ear of the beholder.

Blind studies in which the listener cannot see the instrument being played have shown that even the greatest experts usually cannot discern which is the Stradivarius when the same piece is played by a virtuoso on different violins. Similar experiments with violins made by that other great master instrument-maker, Stradivari's contemporary Giuseppe Guarneri 'del Gesù', revealed the same. Moreover, it seems that violinists themselves cannot tell the difference between playing an old and a new instrument when they are tested under experimental conditions. The fact that a single Stradivarius sold at auction recently for almost £10 million suggests that, secret sauce or not, the instruments remain music to the ears of collectors ... and auction houses.

Of course, it is only the very best, elite modern-day violins – ones that are made with traditional craftsmanship and care combined with modern scientific understanding of acoustics, wood properties, and so on – whose sound is indistinguishable from a Stradivarius. It is not as if you could take any modern violin and expect it to perform as well as the classic. Stradivarius violins remain a remarkable historical achievement: people pay not only for the remarkable quality of performance they allow, but also for the historical value of the instruments.

• • • • • • • • • • • • • •

The Science
Scientific analysis of Stradivarius violins reveals that there is no secret sauce in the wood or the varnish, and that expert listeners and virtuoso players cannot distinguish between high-quality modern instruments and the classic violins.

• • • • • • • • • • • • • •

Find out more
http://www.jeffsextonwrites.com/2011/03/the-stradivarius-myth/

A quick
warm-up on
fitness foolishness

. . . .

DESIGNER EXERCISES CAN 'BURN' FAT FROM SPECIFIC PARTS OF YOUR BODY – Irrespective of how desirable it might be to get rid of excess baggage from the flabbier parts of one's body, there is no way to force the body to shed fat through specific exercises like sit-ups, leg raises, 'bums and tums' training or thigh-busters. With regular exercise and a reduced-calorie diet, fat will be lost from stores throughout the body, gradually and with no specific region responding to particular exercises focused on that area. The good news is that plenty of exercise and a healthy diet will eventually help you get rid of the love handles and the thunder-thighs – and you might even see a six-pack showing through eventually.

. . . .

Find out more

http://oregonstate.edu/recsports/fitness-myths

"

Physical fitness can neither be achieved by wishful thinking nor outright purchase

"

Joseph Pilates

CREAMS CAN BANISH CELLULITE – There are plenty of lotions and potions marketed as being useful for getting rid of cellulite. Unfortunately, for those who part with their hard-earned cash to buy them, these products do not work. Cellulite is not actually a substance in the body; it is not even fat. It is simply dimpling of the skin caused by the irregularities in the fat lying below, the so-called subcutaneous fat. It arises in the same way that depressions appear in a damp cotton handkerchief draped across a cauliflower.

• • • • • • • • • • • •

Find out more

http://goaskalice.columbia.edu/cure-cellulite

WHEN YOU STOP TRAINING, YOUR MUSCLES WILL TURN TO FLAB – Muscles are made mostly of proteins, which are long chains of organic molecules known as amino acids. Fat is made of fat – or, strictly speaking, fatty acids, which are chemically unrelated to amino acids. Muscle proteins cannot be converted to fat, nor fat to proteins. However, if you stop training but maintain your calorie intake you will be eating more food than your body needs. Your body will gradually lay down fat and your weight will increase as a result.

• • • • • • • • • • • •

Find out more

http://www.nytimes.com/2005/07/26/health/nutrition/26real.html

A POT BELLY IS NOT AS BAD FOR YOUR HEALTH AS LOVE HANDLES OR FLABBY THIGHS – Appearances aside, there is mounting evidence that the fat that surrounds our internal organs, visceral fat, is much more dangerous to health than the fat that lies just beneath the skin, the subcutaneous fat responsible for love handles and flabby thighs. That pot belly usually indicates too much visceral fat. Couple that with poor posture and undeveloped abdominal muscles, and there is a tendency for it to all hang out. But it's not just about casting a larger shadow in profile: there is growing evidence from medical research that excess visceral fat – central obesity – is a major risk factor for developing type 2 diabetes, heart disease and the blood vessel problems that lead to a stroke.

• • • • • • • • • • • • •

Find out more

http://healthland.time.com/2012/08/30/can-love-handles-kill-why-having-a-paunch-may-be-worse-than-being-obese/

Does size matter?

The Deceived Wisdom
Herbs, supplements and exercises can increase
the size of the male sexual organ.

. . . .

Most men have accepted how they measure up once
they have reached their peak and left puberty behind:
it's how you know the boys from the men. Sadly for
some, unwarranted feelings of inadequacy can remain, and this
nagging anxiety – an itch that won't be scratched – makes such
worriers the perfect target for the snake oil salesmen of the
modern age: the penis enhancement scammers.

When it comes to penis size, there is probably not a man
on the planet who has not thought about it at some point.
Moreover, there is probably not a single internet user on the
planet who has not received at least one spam email making
claims for products that can increase penile dimensions.

The significance of penis size does seem to be ingrained
in the human psyche: you only have to consider the phallic
symbols that have been depicted, constructed and erected
since prehistoric times. In some primitive sense, the

dimensions of the honourable member may have been the first piece of deceived wisdom, the connection being that a bigger penis equates to greater fertility – which it does not. There may, of course, be a tenuous link between reproductive success if a greater penis size somehow improved the chances of conception, but this does not seem to be the case either. The human penis is larger, relative to body size, than in the other great apes, the chimpanzees and bonobos, gorillas, and orang-utans. Perhaps evolution selected a larger penis in humans as a sexual trait attractive to females in social contexts in which promiscuity and multiple sexual partners were common. After all, the gorilla's penis is much smaller relative to his body size, but he tends to hold a harem of females close and fights off competitors fiercely, so he has no need to impress with it.

So, outside the pornography industry, where the pay cheque might well be commensurate with how an actor measures up during the money shot, is there any need for penis enhancement – and is it even possible? Is there a grain of truth in the endless barrage of spam email offering p3n!s enlargement?

The simple fact of the matter is that surgery is the only way to increase penis size. Even the most enthusiastic surgeons will, however, point out that the procedure is not always successful. It carries with it the risk of constant pain or impotence, and cannot increase the size of the penis by more than a few millimetres, even in a best-case scenario.

According to urologist John F. Bolton, writing in the *British Medical Journal* some time ago, the penis can be made a centimetre or two longer by cutting the penile suspensory ligament that anchors the penis to the pubic arch. This may add a little extra to the length, but cutting that ligament reduces the angle of erection. Boosting the girth is possible by injecting the patient's own adipose tissue (fat) harvested from elsewhere in the body. This is not necessarily a satisfying answer as it can result in an uneven, lumpy appearance.

So, if not surgery, what of the other techniques, pills and potions touted in those deluges of spam messages? Information from clinical trials is limited, and there is no definitive evidence to suggest that any of the methods work at all. Some observers have suggested that using some of the products may have a small placebo effect: they do not boost a man's size, but they can make him feel bigger by boosting his confidence without altering his dimensions, as it were.

Vacuum pumps. These devices claim to increase size by increasing blood flow, but in reality they can damage blood vessels and lead to reduced sensitivity and even impotence. They do have a place in treating clinical impotence, though, in that they can induce an apparent erection.

Pills and potions. There is no known medication that will permanently increase penis size, no pharmaceutical product or herbal remedy that can have any permanent effect on the dimensions of the male organ. Those who believe the hype and seek out these so-called miracle pills online could be simply wasting their money on sugar pills or, worse, taking products contaminated with toxic materials.

Hanging weights. Stretching human tissue usually leads to stretch marks, and there is no evidence that any lengthening will be permanent once the weights are removed. There is evidence of loss of sensitivity, tissue damage and impotence among those who have tried such techniques, however.

Exercises. The penis is not a muscle, so it cannot be made bigger through any form of exercise. Kegel exercises, designed to strengthen the pelvic girdle and focus on the muscles that allow you to hold in urine, even when there is a strong urge to urinate, have potential drawbacks if done to excess, such as retarded ejaculation (an inability to reach orgasm). Although such exercises might appear to allow a man to 'flex' his penis, the penis itself has no muscular tissue that can grow, unlike leg or arm muscles.

So there you have it. The truth about penile enlargement is that it cannot be done safely. More to the point, though, unless he is suffering from the medical condition known as micropenis, no man should worry about it. After all, even the man who feels most inadequate could stand up proud next to a silverback gorilla.

Herbal breast-enhancement

For every million spam emails about penis enhancement, there are probably a dozen or so offering some kind of herbal enhancement for the female breast. Clinical herbalist Guido Masé says that there is no solid, replicated evidence to suggest that any particular herbs could increase the size of the female breast.

There are, however, some reports that herbal preparations known as galactagogues (containing fenugreek or some thistles), which are traditionally used to stimulate milk production after childbirth, might increase breast size to a degree. The size boost is maintained only during lactation, however.

The cosmetic benefits of the herbal remedies that are often used by women wanting bigger breasts without surgery are minimal and short-lived. This does not prevent the industry from making wildly inaccurate claims for herbal enhancers, as anyone with an email account knows only too well.

· · · · · · · · · · · · ·

The Science
Only surgery can permanently increase the size of the penis, and then only by a small proportion. It is just as much a 'phallusy' that size matters: it doesn't, it's what you do with it that counts.

· · · · · · · · · · · · ·

Find out more

http://blogs.webmd.com/mens-health-office/2008/04/
penis-enlargement-myths-and-facts.html

http://www.med.nyu.edu/content?ChunkIID=35539

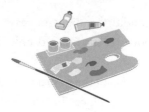

Indigo

The Deceived Wisdom
The seven colours of the rainbow are
red, orange, yellow, green, blue,
indigo and violet.

• • • • • • • • • • • •

In the middle of the seventeenth century, Newton was laying down the law – the law(s) of physics, that is. The apocryphal tale of an apple falling on his head allegedly inspired him to find an explanation for gravity. However, the true inspiration was more likely to have been the many years of planetary observations and the scientific studies he had conducted, rather than a lazy sunny afternoon in his garden at Woolsthorpe Manor. Similarly, he was not out in the garden but inside, in a shaded room, when he used a glass prism to create an artificial rainbow, splitting the white light of the sun into a spectrum. The reverse process was possible if all those colours were recombined by a second prism, though no single colour would produce white light.

Newton fully accepted that the spectrum is a continuous spread of colours, and that there are shades of colour between what we call red and orange, for instance, or green and blue. But Newton fancied that colour might be harmonious, like the Western musical scale and, following some other-worldly structure, should have seven specific colours: red, orange, yellow, green, blue, indigo and violet. Orange and indigo, Newton suggested, might be like semitones between two notes of the major scale. There is, however, evidence that Newton labelled his colours not on the basis of their appearance in spectrally split white light, but from an examination of the palettes of contemporary painters. Pigments and paints are not light, and behave very differently. Mix all the 'colours of the rainbow' on a palette and you will get a brown-black smudge, but mix the same colours of light and you produce white.

Our modern understanding of the visible spectrum is that it is merely the continuous spread of electromagnetic radiation to which our eyes happen to be sensitive. Beyond the red is infrared (a form of heat), then microwaves and radio waves, all tending towards lower and lower energy. At the other end, beyond violet is (obviously enough) ultraviolet, then X-rays and gamma rays, all higher-energy forms of electro-magnetic radiation.

The US National Bureau of Standards correlates particular wavelength bands with a specific colour: 400–465 nanometres (nm) is violet, 465–482 nm is blue, 482–487 nm greenish blue … 597–617 nm reddish orange and 617–780 nm red. There is no space for indigo. Of course, orangey red or reddish orange are not definitive descriptions. You might just as well refer to the colour between yellow and green not as greenish yellow but as 'chartreuse' – but only if that word is in your vocabulary. It's not so much that a particular colour is present or missing from the rainbow, but that there are as many colours as you care to perceive, and that very much depends on your culture, your understanding of art and physics, and your eyesight.

• • • • • • • • • • • • •

The Science
The colour indigo was 'invented' by
Sir Isaac Newton to base his theory of light on
a more pleasing number, seven, which has many
mystical and musical allusions. But the rainbow
spectrum is a continuous spread of colours – of
which there are as many as there are
words available to describe them.

• • • • • • • • • • • • •

Find out more

http://en.wikipedia.org/wiki/Rainbow#Spectrum

Health hokum

• • • • • • • • • • • •

SOME FORMS OF FIRE-RETARDANT ASBESTOS ARE
SAFE – Asbestos is a fibrous mineral composed of sili-
cates and was once widely used as a construction material because it could be formed into board with fire-retardant
properties. Although it was used from the nineteenth century
until well into the twentieth century, it was recognized early on
that it can cause health problems, particularly for those mining
the material and working with it in industry and construction.
The fibres of all forms of asbestos – white, brown and blue –
can, if inhaled, become lodged deep within the tissues of the
lungs. The fibres are thought to cause chronic inflammation,
and this can ultimately lead to lung cancer, though scientists
are yet to work out how the physical presence of these fibres
causes chemical changes in lung cells.

All forms of asbestos can cause cancer, asbestosis or
malignant mesothelioma. Despite the risks, only fifty-two
nations have banned asbestos outright.

• • • • • • • • • • • •

Find out more

http://www.cancer.gov/cancertopics/factsheet/Risk/asbestos

CRACKING YOUR KNUCKLES WILL GIVE YOU OSTEO-ARTHRITIS – Concerned and vexed parents will often warn knuckle-cracking children that their irritating habit will lead to painful and potentially debilitating osteoarthritis in later life. However, several research studies have failed to demonstrate any link between the habit and the development of problems in the finger joints.

Most recently, American scientist Kevin DeWeber and his colleagues showed that among hundreds of people tested repeatedly and even long-term, knuckle cracking had no correlation with whether or not they developed osteoarthritis. This seems like an obvious finding, because the 'cracking' is simply the popping sound of dissolved gases quickly released from the natural lubricating fluid present in the membranes surrounding the joints, but their work helps to provide the scientific evidence that knuckle cracking will not lead to osteoarthritis.

· · · · · · · · · · · ·

Find out more

http://www.personal.psu.edu/afr3/blogs/SIOW/2011/09/
knuckle-cracking-myth-is-cracked.html

YOU NEED TO DRINK EIGHT TO TEN GLASSES OF WATER EVERY DAY TO STAY HEALTHY – Staying properly hydrated is important for health, but the suggestion that everyone must drink at least two litres of water every day is misguided health hokum. Our bodies are almost two-thirds water and deviate very little from that level unless we are in an extreme environment of heat with no food or drink. However, there is a huge difference between feeling thirsty and being dehydrated.

"

There are in fact two things, science and opinion; the former begets knowledge, the latter ignorance

"

Hippocrates

If you eat regular food and partake of the odd cup of tea, coffee, fruit juice or even water during the day, you will consume perfectly adequate amounts of water from the food and drink for good health. Indeed, one litre of water from any source is enough to replace fluid lost through sweating and urination, and most estimates suggest that you could obtain that much from your food alone. Interestingly, even though coffee contains caffeine, which is a diuretic and makes you produce urine, unless you are drinking nothing but strong espresso the water from the drink will usually compensate for any increased loss through urination. Of course, when you sweat as a result of exertion (whether exercise or physical labour) you will need to increase the amount of water you take in. Sports scientists estimate that you need to drink one litre of water for every 1,000 kilocalories of exercise. For comparison, an hour's cycling requires approximately 500 kilocalories of energy, so that's an extra half a litre of water you'd need to drink.

The precise quantities depend on your thirst, which is usually the best indicator of whether or not you need a drink, not a prescription in a health or lifestyle magazine.

· · · · · · · · · · · · ·

Find out more

http://www.snopes.com/medical/myths/8glasses.asp

Only dedicated practice makes perfect

The Deceived Wisdom
Practice makes perfect.

• • • • • • • • • • • •

Being exceptional at something is often attributed to one's genetics. Talent is passed down from parents or grandparents, it seems, whether it is musical or artistic skill, ability with numbers or being great at juggling. No doubt there are significant genetic factors involved, but there are almost certainly environmental factors in the mix too. Perhaps the two work together, one boosting the other, so that those remarkable genes give rise to remarkable talent only if the skills are suitably nurtured.

However, many people now recognize that talent is learned and earned through extended and intense practice of a skill. Genes, they say, have little to do with it. This idea is encapsulated in a golden rule made popular by the writer Malcolm Gladwell in his book *Outliers*. This '10,000 hours of practice' rule is based on research by psychologist Anders Ericsson of Florida State University. Apparently, the rule tells us, a mere 10,000 hours of dedicated practice in your particular field is sufficient to bring out the best in you.

In essence, Ericsson's theory suggests that sufficient practice in a particular skill can take anyone to the level of proficiency equivalent to that heard in the playing of a top concert pianist. Gladwell embraces this idea, pointing out that all great sportspeople, performers and even computer programmers apparently got in their 10,000 hours of practice in their particular art early in life. This helped them to excel early on, allowing them to shine while their less diligent contemporaries were still grappling with the basics.

Gladwell cites the figure of 10,000 hours in connection with the well-known period in the early musical careers of The Beatles when they played almost endless nights in the nightclubs and bars of Hamburg in Germany between 1960 and 1964. This opportunity gave them something few musicians had during that era – plenty of time to practise. Ultimately, says Gladwell, this is what made the Fab Four top musicians and songwriters.

He also cites Bill Gates, the founder of computer software giant Microsoft, famous for the Windows operating system for personal computers and Office software. Gates is apparently a great example of the 10,000-hour rule. He had access to a computer in 1968 at the age of 13, at a time when most of his school friends would have been playing baseball or dreaming of putting flowers in their hair and heading to San Francisco. This gave him a substantial head start in the area of computer programming and apparently allowed him to build his company at a much younger age than he might otherwise have been able to.

Many of us imagine that hours and hours spent on our chosen pursuit are somehow getting us towards that target of 10,000. I've played guitar since the age of 12, but I don't imagine that

I'm anything but a total amateur musically speaking – I've not put in the dedicated, repetitive practice. Anyone who has heard me strumming might suggest that I plug headphones into my guitar amp and practise for another 10,000 hours before letting anyone ever hear me play again. Ericsson, the psychologist on whose research Gladwell apparently based his interpretation of the 10,000-hour rule, might well agree. Not because he has heard me play, but because that rule is not quite as it may seem.

To notch up 10,000 hours would require about 90 minutes' practice every day for 20 years. This might explain why the typical child learning the piano will never make it to concert level. Three hours a day gets you there within a decade, so start at the age of 10 and you're done before you're out of your teens. Unfortunately, the moment the 10,000-hour mark is reached is not a skills tipping point. Learning and gaining expertise are gradual processes; skills evolve slowly, with practice. And there is a vast range of time periods over which different individuals reach their own peak of proficiency – their concert level, you might say – in whatever field.

Ericsson is on record as emphasizing that not just any old practice counts towards those 10,000 hours. It has to be dedicated time spent focusing on improvement. Not all the examples in Gladwell's book qualify as such deliberate practice: writing computer programs and playing ice-hockey matches, for instance, may not count. It is not a matter of simply taking part in an activity, Ericsson argues. More to the point, for musicians it seems that the winners of international competitions are those who have put in something like 25,000 hours of dedicated, solitary practice – that's three hours' practice every day for more than 20 years. For sportspeople, there are physical limits on how much dedicated practice is possible.

The question of whether or not 10,000 or even 25,000 hours of practice is enough does not tell us anything about whether some people are born with a particular talent. We do not yet know whether anyone with strong enough motivation and the spare time could become a virtuoso simply through deliberate practice, year in year out.

Scientifically speaking, 10,000 hours is not a precise figure but shorthand for 'lots of dedicated practice'. If you want to achieve concert level, whether at the piano or on the running track, 10,000 hours is a good starting point. Double that and you might even start winning international competitions. But however you look at it, being the best requires a lot of time and effort, and not many people want to dedicate so much of their lives to a single pursuit. So while practice might get some of us close to perfection, for many of us it is probably an unattainable goal. That is no reason not to give it a try, of course. Some day I might even unplug those headphones once more.

· · · · · · · · · · · · ·

The Science
Most people do a relatively limited amount
of practice to 'perfect' their art, whether wielding
a cricket bat, playing chess, singing or programming
computers. It can make you perfectly competent –
but not necessarily excellent. Even 10,000 hours
of dedicated practice may not be enough to
give you the skills of a virtuoso.

· · · · · · · · · · · · ·

Find out more
http://aubreydaniels.com/pmezine/expert-performance-apologies-dr-ericsson-it-not-10000-hours-deliberate-practice

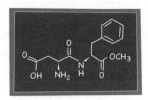

Sweetener
for my sweet

The Deceived Wisdom
Diet cola containing the artificial sweetener
aspartame can cause blindness.

• • • • • • • • • • • •

For years, the artificial sweetener saccharin, beloved of the calorie-counters who might choose a diet drink to make up for the consumption of doughnuts, burgers and fries, was the poster child of the anti-chemical movement.

Synthetic and sickly sweet, saccharin soon became anathema in some quarters when laboratory evidence began to emerge that it might cause cancer. Laboratory tests can be deceiving, however. The useful but notoriously simplistic Ames test for cancer-causing substances can be used to demonstrate that the most innocuous of chemicals – water, even – can cause changes in DNA that might be linked to cancer growth. Nevertheless, the harm was done. Saccharin has since been scratched from food and drink ingredients lists in several parts of the world, even though it was only ever used in tiny quantities and helped to preclude the tooth decay and weight gain associated with overindulging in sugary foods and drinks. The search for alternative artificial sweeteners with none of the 'empty calories' of sugar was already well under way when things turned bitter.

One of those alternatives was aspartame. Whereas zero-calorie saccharin is a synthetic molecule originally derived from coal tar in the nineteenth century, and with a chemical structure that somehow stimulates the tongue into believing it is tasting something sweet, aspartame is a whole lot more 'natural'. This compound also goes by the formal name of N-(L-alpha-aspartyl)-L-phenylalanine, 1-methyl ester. But this worryingly complex Greek-sounding label belies the fact that aspartame is composed of the very building blocks of life from which all proteins are made: amino acids. Let's take a closer look at that name. The aspartyl bit refers to the amino acid aspartic acid, and the phenylalanine is another amino acid. The L's tell us that these are the natural forms that can be digested by the body, just like any amino acid from our meat and two veg (or fish, pulses, nuts and other protein-rich foods). The N and the alpha tell us about how the two amino acids are joined together to form aspartame itself.

Despite its innocuous constitution, the artificial sweetener aspartame has become the target of many a scaremongering tabloid story or blame-seeking activist agenda. There is even a Facebook page dedicated to getting it banned. Claims have been made that it causes problems ranging from muscle cramps and seizures to headaches and depression … even Gulf War Syndrome, a collective term for various problems faced by veterans of the 1991 war fought by the US and its allies against Iraq. These claims have been endlessly repeated in email chain letters and poorly researched tabloid stories.

But aspartame's bad reputation is unjustified. There is only one valid concern. Babies born with the rare genetic disorder phenylketonuria (PKU) cannot process the amino acid phenylalanine and so must avoid any food with proteins rich in that particular amino acid as they grow into toddlers. If they don't, the breakdown product of phenylalanine builds up and stops normal brain development. A person with PKU faces a lifetime of dietary restrictions: meat, fish, eggs, nuts, cheese, legumes, milk and other dairy products are permanently off the menu. Diet drinks containing aspartame are the least of their concerns given those dietary restrictions, but they too must be avoided.

For most of us, phenylalanine is not a problem and we are free to eat all those proteins and drink diet drinks. In the body, aspartame is broken down into its component amino acids which are then processed in just the same way that they would be if they were ingested from meat, dairy products or any other protein. The chemical bridge used to hook the pair of amino acids together also means that a tiny amount of methanol is released in the process.

Could it be that methanol is what those wary of aspartame are so scared of? After all, in a large enough dose or with chronic exposure, methanol can cause permanent blindness. Any doctor will testify to the sight problems they find in people drinking methylated spirits as a cheap but toxic alternative to alcohol. Methanol's fatal dose can be as low as 25 millilitres, about five teaspoons, but is usually well over 100 millilitres – the equivalent of a teacup of pure methanol. Could people drinking diet cola, for instance, be exposed to toxic doses of methanol when the aspartame is broken down in their body?

A standard can of diet cola contains up to about 180 milligrams of aspartame. When that quantity is all broken down and absorbed by the body, it produces 20 milligrams of methanol, which is about 0.025 millilitres. That is about one-thousandth of the lowest known toxic dose. To put it another way, to be exposed to a dose that could damage your sight you would have to drink 400 cans of diet cola. A day. Even the most hardened dieter would not be able to drink that much in a misguided quest to lose weight. But is there a risk that drinking two or three cans a day might lead to enough methanol accumulating in the body to cause problems within a few months?

Methanol is a cumulative poison, which is why people who drink methanol-contaminated hooch gradually succumb to its blinding toxicity. However, if all the methanol released from ingested aspartame is absorbed, it would take just an hour or so for it to reach its peak concentration in the blood. The liver gradually breaks down methanol into formaldehyde, which is then converted into formic acid, the substance squirted by stinging ants. It is this acid rather than the methanol itself that blinds the meths drinker. Fortunately, after about 90 minutes half of the methanol will have been broken down and the formic acid excreted by the kidneys. After another 90 minutes, half again will have gone, and so on until after a day or so almost all of the methanol will have left the body.

Drink a single diet drink and you would, as mentioned, ingest 0.025 millilitres of methanol. Drinking a six-pack of diet cola in 90 minutes – a difficult task – would mean you had ingested about 0.15 millilitres, which is a little more than 1 per cent of the minimal harmful dose of methanol. But after an hour and a half the dose present in your body will have been broken down by the liver to 0.075 millilitres. Drink another six-pack of diet cola during the next hour and a half and you would ingest another 0.15 millilitres, bringing your total up to about 0.26 millilitres. That quantity will halve in 90 minutes. If you were physically able to drink a six-pack of diet cola every 90 minutes all day long, the maximum amount of methanol that could accumulate would be just under 0.3 millilitres. That is a tiny fraction, just 3 per cent, of the minimal dose known to cause vision problems. Compare that with a person drinking methylated spirits, which is one-tenth methanol and the rest ethanol, water and other components. A litre of meths contains about 100 millilitres of methanol – that's 40,000 times as much methanol as might theoretically be absorbed after drinking a can of diet cola.

The scaremongering about aspartame seems to focus on the breakdown products. Most peculiarly, there have been claims that diet drinks exposed to high temperatures are somehow more dangerous, which is partly why they were blamed for Gulf War Syndrome. But it would make no difference to their effects whether the aspartame was already broken down in the can or broken down after the drink had been drunk. Anyway, once it is broken down, aspartame no longer tastes sweet, and a degraded diet cola would have a very nasty taste without its sweetener.

So, if you're having sugar-free drinks as part of a calorie-controlled diet, the last thing you should worry about is whether or not the artificial sweetener is somehow bad for your health and your eyes. It's the sugar-laden doughnuts you need to worry about.

· · · · · · · · · · · ·

The Science
Aspartame is composed of two entirely natural amino acids which are found in meat, dairy products and vegetables. It is these amino acids that are released and processed by the body. A tiny

amount of methanol might also be produced during digestion, but you would have to drink hundreds if not thousands of cans of diet cola every day to accumulate a toxic dose of methanol.

• • • • • • • • • • • •

Find out more

http://www.sciencemediacentre.co.nz/2008/09/
12/aspartame-is-it-safe/

Another miscellany of misconceptions

. . . .

WE USE ONLY 10 PER CENT OF OUR BRAIN – Aside from how silly this is, given all the different functions of the brain involved in thinking, breathing, digesting food, moving muscles and generally keeping our bodies alive and working, modern brain scans clearly show that every part of the brain is used at some time or another. The brain has even been shown to be highly active throughout both of its hemispheres when we are asleep.

. . . .

Find out more

http://health.howstuffworks.com/human-body/systems/
nervous-system/ten-percent-of-brain.htm

ADDING SALT WILL CUT COOKING TIMES – Adding a solute such as common salt to a solvent such as the water you use to boil your Brussels sprouts will raise the boiling point. This means that salted water boils at a temperature above 100°C, the normal boiling point of water at atmospheric pressure. Unfortunately, the rise is only a small fraction of a degree, and this has very little impact on the time it takes to cook food. Adding salt is more about improving flavour than reducing cooking times. Indeed, sodium chloride (common table salt) can mask any bitter tastes in the food, making those boiled sprouts even tastier.

• • • • • • • • • • • •

Find out more

http://lifehacker.com/5847591/
10-stubborn-food-myths-that-just-wont-die

CHEMICAL-FREE PRODUCTS ARE BETTER FOR YOU – There is no such thing as 'chemical-free': everything you can touch (and everything you can't) is made of chemicals. Water is a chemical, as is the oxygen in the air we breathe, the cellulose in grass, the sand on the beach, your fingernails, your eyelashes, your brain. Moreover, 'natural' does not equate to 'safe': think snake venom, deadly nightshade, jellyfish stings and asbestos.

• • • • • • • • • • • •

Find out more

http://www.senseaboutscience.org/data/files/resources/5/
MakingSenseofChemicalStories_July08-Reprint.pdf

THERE IS NOTHING IN A VACUUM – The laws of physics suggest otherwise. Even if you suck all of the gas out of a container, there are still quantum fluctuations in the vacuum in

"

The most exciting phrase to hear in science, the one that heralds the most discoveries, is not 'Eureka!' (I found it!) but 'That's funny...'

"

Isaac Asimov

which subatomic particles and their antiparticles constantly pop in and out of existence.

• • • • • • • • • • • • •

Find out more

http://www.chalmers.se/en/news/Pages/
Chalmers-scientists-create-light-from-vacuum.aspx

LIGHTNING NEVER STRIKES THE SAME SPOT TWICE

– The maintenance team at the Empire State Building in New York City will testify that their building is struck by lightning at least twenty-five times each year. Other high points on the landscape suffer an equally discouraging succession of bolts from the skies, and there are several reports of people who have survived multiple strikes. Worldwide, there are about thirty lightning strikes that reach the ground every second. It is perhaps therefore not surprising that lightning strikes the same place more than twice quite often.

• • • • • • • • • • • • •

Find out more

http://www.nasa.gov/centers/goddard/news/
topstory/2003/0107lightning.html

APPLE MACS DON'T GET COMPUTER VIRUSES – In fact,

there are lots of viruses and other forms of malicious software – malware – that infect Apple products. It is, however, commonly believed that Apple machines are immune, partly through clever marketing over the years but also because, historically, their security strengths have made them a harder target for writers of malicious software. Moreover, there are many more Microsoft Windows machines in use than Apple Macs, so when a piece of malware infects a Windows machine it has more chance of spreading – thus raising the profile of malware and denigrating the reputation of Windows.

* * * * * * * * * * * *

Find out more

http://blogs.mcafee.com/consumer/myth-apple-products-
don%E2%80%99t-get-viruses

YOUR MOBILE PHONE WILL FRY YOUR BRAIN – Many people worry that their mobile phone will give them brain damage or cancer. There is absolutely no evidence for this. In two decades of studies, says the World Health Organization, 'to date, no adverse health effects have been established as being caused by mobile phone use'.

* * * * * * * * * * * *

Find out more

http://www.hpa.org.uk/web/HPAweb&HPAwebStandard/
HPAweb_C/1317133823488

YOU COULD DROWN IF YOU SWIM RIGHT AFTER EATING – Although the gut does divert blood to help with digestion, there is no evidence that this has ever affected anyone's ability to swim, and there is no valid official recommendation on not eating before swimming from any sports or safety organization. Of course, some people find exercising after eating uncomfortable, but it is not physically dangerous. Children will often jump down from their place at the dinner table and be running, swimming, bouncing and jumping without any ill effects. Swimming after drinking alcohol is a different matter ... and not something that is encouraged for children, or adults.

∙ ∙ ∙ ∙ ∙ ∙ ∙ ∙ ∙ ∙ ∙ ∙

Find out more

http://www.snopes.com/oldwives/hourwait.asp

MOST OF THE BODY HEAT YOU LOSE IS THROUGH THE TOP OF YOUR HEAD

– If you are brave enough to go outside naked on a cold day, the rate of heat loss from any part of your body will be approximately the same from any given area. However, on cold days we tend to wrap up warm with trousers, tops, shirts, pullovers, coats and woolly socks. If you don't add a hat to the mix, then the rate of loss of heat from your head will, of course, be higher than from your other covered body parts, but put on a thick, insulating hat and you will stay cosy and warm. Incidentally, going outside with wet hair on a cold day will not increase the risk of your catching a cold. Exposure to the cold-causing virus is the only thing that increases that risk.

∙ ∙ ∙ ∙ ∙ ∙ ∙ ∙ ∙ ∙ ∙ ∙

Find out more

http://health.howstuffworks.com/diseases-conditions/
cold-flu/wet-head-cold.htm

Six degrees

The Deceived Wisdom
We are all connected to each other by a
maximum of six degrees of separation.

. . . .

We have almost all experienced the weird coincidence
of bumping into a stranger and learning, to our
mutual amazement, that we have mutual friends or
acquaintances or colleagues who know co-workers. It happened to me when I was working in the USA. I met someone
who knew several of my university friends and who later, quite
by chance, met my sister-in-law, all independently.

Such experiences suggest that the 'small world' phenomenon
described by psychologist Stanley Milgram may have some
basis in reality. It was Guglielmo Marconi, developer of
radio communication, who in his 1909 Nobel Prize-winner's
speech put a number on this phenomenon, suggesting that
each person in the world would be connected by an average
of 5.83 interpersonal connections – hence the 'six degrees of
separation'. Of course, when averages are mentioned there are
always above-average values and ones below.

Well-connected jet-setters working in the media in densely

populated metropolises might have much shorter connections to others. Indeed, a much-loved game emerged from the possibility of connecting popular movie actor Kevin Bacon to other actors (the game was called, somewhat predictably, 'Six Degrees of Kevin Bacon'). It is usually possible to find a chain linking him to almost any other actor via his movies. However, the phenomenon seems not to be unique to Bacon, and the linking game also works with many other actors. Scientists and mathematicians play a similar game with their published research papers, giving themselves a so-called Erdős number that places a value on the 'collaborative distance' between papers to which they have contributed and the interconnections leading back to the prolific Hungarian mathematician and author Paul Erdős.

With the advent of the World Wide Web, and in particular social networking sites such as Facebook and Twitter, the chances of finding connections between people have grown considerably. Some people at the 'hubs' or nodes of the network are better connected; those out on a remote limb may require a far more than the average six links to reach others. Milgram's experiments in the 1960s and 1970s that attempted to prove his 'small world' theory do not bear close scrutiny and were not particularly successful, given the number of failures in connecting people.

Yet even if we were all connected to one another by a short chain of six contacts, there is a theory – well supported by research in anthropology – that our brains are hard-wired by evolution to cope with no more than about 150 close contacts. That's 150 people you know and 'love'. The theory may explain why hunter-gatherer villages have a maximum population size of about 200 to 250; why certain social groups, such as the Hutterite religious sect, split their communities once they reach this size, and have done for centuries; and maybe even why fighting groups work best when they have fewer than 200 members. We simply haven't got the brainpower to really care about more than that number of people, because we can't keep track of all the relationships between the group members if there are more. There is even research suggesting that companies and organizations of this approximate size can work better and more efficiently than much bigger companies. Indeed, very large companies tend to be divided into smaller divisions or spin-off companies created for this and other reasons.

This ideal number, commonly known as Dunbar's number after British anthropologist Robin Dunbar, does not have a fixed value, and other researchers have found that it could be much higher or even lower than the average suggested by Dunbar. What is clear is that, though we might somehow find a way to connect with almost every one of the earth's population of 7 billion people, we are only likely to really care about a couple of hundred or so at most, regardless of whether they have been in a movie with Kevin Bacon.

. . . .

The Science

There is little evidence that we are connected to everyone else by just six connections – and even if we were, the human brain has evolved to cope only with a maximum number of relationships.

. . . .

Find out more

http://www.psychologytoday.com/articles/200203/
six-degrees-urban-myth

Round 2 of
fitness foolishness

• • • • • • • • • • • •

YOU CAN 'SCULPT' MUSCLES BY VARYING
RESISTANCE EXERCISES – Muscles have specific
points of attachment connecting them to your skeleton.
They flex by contracting from these points and relax when the
resistance is removed. It makes little, if any, difference at which
angle you flex them with weights, ropes and pulleys, or some
other gym machine. Variations on an exercise may well cause
neighbouring muscles to contract, for example if you hold a
barbell at a different angle than usual or dangle yourself from a
weights bench, but sculpting the original muscle by working it
in different ways is impossible.

• • • • • • • • • • • •

Find out more

http://www.askthetrainer.com/what-is-muscle-tone/

WEIGHTLIFTING WILL MAKE YOU BIG AND BULKY –
Anyone who has tried to bulk up will soon discover the deception in this fitness fallacy. It is partly true for men, but to gain

muscular mass you must also eat more food – carbohydrates, fats and protein. Women face the additional problem of usually not having adequate levels of testosterone, the male sex hormone that stimulates muscle growth.

• • • • • • • • • • • •

Find out more

http://www.fitnessgoop.com/2011/05/debunking-the-%E2%80%9Cbig-and-bulky%E2%80%9D-strength-training-myth/

THERE IS MORE TO HOMEOPATHY THAN SUGAR PILLS AND WATER – Homeopathy involves repeated dilution of the 'active' ingredient. A common homeopathic method of preparation is to add a small amount of this ingredient to a certain volume of water and dissolving. A single drop of this solution is then added to the same volume of water as before, a drop of this further diluted solution is then added to the same volume of water, and so on. This can be continued ten, twenty, thirty, fifty or even a hundred times.

Let's assume that the homeopath starts by dissolving 1 mole of the active ingredient in a litre of water. A mole of a substance is a measure used in chemistry, and is equivalent to 6.02×10^{23} molecules – which sounds a lot. But after a generous tenfold dilution carried out 24 times in succession, there will remain a physically impossible 0.6 molecules of the original ingredient in the solution. Given that homeopaths often dilute their ingredients a hundred times or more, there is effectively zero chance of even a single molecule of the active ingredient remaining in the prescription offered.

Homeopaths have claimed that water retains a 'memory' of the original molecule through all these repeated dilutions.

Of course water does not retain a 'memory' of the original substance: water is a liquid, and its trillions upon trillions of molecules are in constant jittering movement, linking up for billionths of a second through temporary hydrogen bonds and then separating and forming new links. The principle of homeopathy is akin to burying footballs in a desert sand dune, allowing a hundred sandstorms to take place and expecting then to find football-shaped hollows in the dune.

Homeopathy is at best an expensive placebo. When it is used as an alternative to proper medicine, as an antimalarial drug or for treating cancer or AIDS, for example, all it does is give patients false hope.

• • • • • • • • • • • •

Find out more

http://theconversation.edu.au/doctors-orders-debunking-homeopathy-once-and-for-all-1393

DETOXING IS GOOD FOR YOUR HEALTH – You cannot choose to detoxify your body: everything you eat releases some toxins, also known as natural poisons. Your liver will process these and allow them to be excreted in urine. Indeed, the very metabolic processes that keep you alive produce waste products. The kidneys will usually excrete these too, while the waste carbon dioxide you produce is expelled by the lungs. Even if you drink only pure water and eat nothing, your living cells will continue to generate natural free radicals and oxidizing compounds, and ultimately these too will either break down or be processed and excreted.

In addition, there is nothing you could ingest that would 'detox' your body as that substance would also be digested, release its own waste products, and stimulate metabolic processes that would require the liver and kidneys to process them. Compounds supposed to have 'detox' properties have been tested to see whether they reduce the effects of caffeine after drinking coffee, for instance by boosting enzymes that break down coffee. But you have to take the substance for several days before drinking the caffeine, and more to the point, it blocks the action of the caffeine almost immediately. If you are worried about caffeine, it's easier just to drink decaf.

The true deceived wisdom about detox, however, is that somehow our modern lifestyle makes our insides congested and dirty – so dirty that we need to clean them by 'detoxing'. It is wise to seek medical treatment urgently if you ingest or breathe in something poisonous, and it is a good idea to take everything, including specific foods and alcohol, in moderation. But dispense with the detox: it does nothing for your health.

* * * * * * * * * * * * *

Find out more

http://www.senseaboutscience.org/pages/debunking-detox.html

Red, red wine

The Deceived Wisdom
The cork must be removed from a bottle of red wine well
in advance of drinking to allow the wine to 'breathe'.

. . . .

A ny good dinner party host knows that the corks must be
pulled on the bottles of red wine well ahead of the meal
so that the wine can 'breathe'. Any good wine waiter
knows that such an affectation is a pointless act and has no effect on boosting the body, improving the bouquet or removing
any bitter aftertaste. A good red wine will taste fine, provided
it is not too cold, even if the cork is removed just before it is
poured and quaffed.

What the good sommelier also knows is that sometimes
a less than adequate red wine can be improved. This is done
not by simply opening the bottle but by pouring it quickly into
another container, a decanter. Under pressure, a sommelier
might explain this little trade secret as aiding aeration to
'smooth some of the harsher aspects of the wine'.

However, according to Andrew Waterhouse, Professor of
Enology at the University of California at Davis, even quickly

decanting and leaving the wine to stand for half an hour is not long enough for the astringent tannins in younger and less pricey wines to be broken down. Tannins give 'dryness' to a wine, but make you pucker up if the content is too high. Waterhouse explains that it takes at least a day for these natural chemicals to be broken down. Unfortunately, if you leave a bottle of wine to breathe for that length of time, not only will your guests have broken down and departed but bacteria, known as acetobacter, will have invaded the wine and started to convert the alcohol content to acetic acid, or vinegar.

As for the idea that a wine becomes smoother over the course of a meal, that is more down to your taste buds and nose than anything in the wine changing chemically. After the first few mouthfuls the wine may seem smoother, but what's happened is that your mouth has become more accepting of the wine's taste and its level of astringency. As to the evaporation of malodorous compounds, these are rare in modern winemaking. If a bottle is 'corked' it will have a particularly unpleasant, musty odour caused by the presence of TCA, 2,4,6-trichloroanisole. This compound is harmless in the quantities present in a corked wine, but it makes the drink unpalatable.

For white wine there is every reason *not* to let it 'breathe'. White wines have little tannin content, and the fruity aromas that often accompany them are usually due to volatile compounds in the wine. Decanting a bottle of white wine will cause some of this volatile content to be lost, and the wine will lose much of the bouquet that the winemaker hoped you would experience. The same sad loss will happen if the wine is warm or allowed to become warm once it has been opened. If you ever see a wine waiter decanting a bottle of white wine, it's

"
Wine is at the head of all medicines; where wine is lacking, drugs are necessary

"

Talmud

time to decant yourself from the restaurant and eat elsewhere.

There's one more piece of deceived wisdom about red wine. You often hear claims that red wine contains some kind of medical panacea, whether it is the antioxidant compound resveratrol or one of countless other natural chemicals found in the drink. Certainly, wine contains lots of natural chemicals. One of them is ethanol, which we usually refer to as alcohol. Any marginal benefits of antioxidants or other health-giving compounds in wine are cancelled out many times over by the presence of ethanol, which despite our enthusiasm for drinking it is a toxic chemical. Scientific studies from around the world show that more than half of all alcohol-related deaths are due to heart disease, cancer or liver damage. So much for the red wine panacea.

· · · · · · · · · · · ·

The Science

The surface area of wine exposed to the atmosphere by removing the cork or unscrewing the cap from a bottle is far too small to allow any significant exchange of gases with the liquid. Simply swirling the wine in the glass once it is poured has little effect, but the sommelier's secret is that decanting a lower-quality wine or a younger wine that still needs to age can improve it slightly by allowing some malodorous compounds in the wine to evaporate.

· · · · · · · · · · · ·

Find out more

http://wineserver.ucdavis.edu/people/faculty.php?id=14

The moon's not a balloon

The Deceived Wisdom
The full moon looks bigger when it is closer to
the horizon than when it is high in the sky.

• • • • • • • • • • • •

To some people, the full moon appears larger in the sky when it is nearer the horizon than when it is high in the sky. This is the so-called 'moon illusion', and works the same with a sunrise or a sunset. Generations of artists have exploited it in their paintings, but it has puzzled psychologists for many years.

The fact is, however, that on any given day or night the full moon and the sun both appear about half a degree wide. (Imagine a huge circle tracing out 360 degrees all around the sky – the sun and moon both take up about 1/720 of its circumference.) This coincidence of apparent size is why total solar eclipses are possible, as the moon's disk can completely cover the sun's, regardless of where in the sky the eclipse appears to be. There is no weird atmospheric refraction or magnification effect taking place, nor any mystical phenomenon occurring.

Subjectively, the sun and moon can both seem as much as one and a half times bigger at the horizon than when they are high in the sky. Artists of all ages often represent a low-lying moon as being a lot bigger, and commonly show the rising or setting sun as spanning a much bigger portion of sky than it actually does. Examples of gross distortions of the sun or moon's size in art can be found in Vincent van Gogh's *Sower with Setting Sun* (ten times too big), Honoré Daumier's *The Bluestocking* (four times), Samuel Palmer's *Coming from Evening Church* (five times) and Ernest Briggs's *The Northern Twilight – Returning from the Fishing* (four times).

There are many explanations of the moon illusion, but none are entirely satisfactory. Among them is the notion that we perceive the sky as a flattened dome, so that the horizon is from our internal perspective farther away than the zenith directly above. An object lying near the horizon thus appears bigger because our brains subconsciously scale it for distance. Think of it this way: scattered clouds across the sky give the odd impression of a gently curving sky as we look above us and then lower our gaze towards the horizon. It's perhaps all about clues to distance, which in the sky are few and far between.

The problem with this explanation is that, given that perception, the moon should appear bigger but farther away at the horizon, but to most people it doesn't – it appears bigger and closer. Psychologists Helen Ross and Adele Cowie tried to find an explanation by testing the moon illusion with children aged 4 to 12 and adults aged 21. Most people in their tests drew the moon bigger in a sketch or painting of the moon near the horizon, but smaller if they drew the moon higher in the sky.

For objects that are fairly close by, we can estimate their size based on everyday experience: for example how they appear to move and by how much, and comparison with their surroundings. For more distant objects our estimates are imperfect, and we have to somehow take into account the distance to an object in working out how big it is.

Of course, the true distance to the moon is beyond our everyday experience, so our estimates of its size are somehow distorted. The low-lying moon might be framed by silhouetted trees or buildings, whereas the stars are the only reference points for a moon high in the sky. Weirdly, the moon illusion seems to vanish if we tilt our head or bend over to look at the moon. Our perception changes depending on the angle at which we look at something relative to the local terrain. How this effect helps us understand the moon illusion is not entirely clear. But what is clear is that the moon illusion is real to some people, though it is a purely psychological effect and not a physical phenomenon. Nevertheless, without it, atmospheric paintings of the nocturnal horizon would be far less dramatic.

All other things being equal, the moon is actually farther away when we see it 'close to the horizon' rather than overhead – by about one earth radius –so at such times its size in the sky is ever so slightly smaller, not bigger. Of course, the moon's apparent size in the sky does vary ever so slightly, because of its elliptical orbit (though that doesn't correlate with where we see it in the sky).

. . . .

The Science
The area of the sky covered by the full moon is pretty much the same regardless of its position in the sky on a given night; moreover, it is never physically bigger in the sky when it is close to the horizon.

. . . .

Find out more
http://facstaff.uww.edu/mccreadd/

Facts and
non-fiction

. . . .

HOUSEHOLD DUST IS MOSTLY SKIN – The precise composition of house dust depends on where you live, but it can be a mixture of dust from building materials, sand, soil, clothing fibres, plant fibres, fragments of hair (animal and human), pollen and insect faeces. Only a tiny fraction is formed from flakes of human skin.

TEFLON WAS A SPIN-OFF OF THE SPACE RACE – The non-stick plastic polytetrafluoroethylene (Teflon) was invented serendipitously in 1938 by Roy Plunkett of Kinetic Chemicals in New Jersey. It was used for valves and seals and for 'non-stick' cooking utensils long before the space race got going.

A WATCHED POT NEVER BOILS – While this is more of an aphorism about patience, it is patently untrue and would defy the known laws of physics if it were not. However, at the quantum level the very observation of subatomic particles does interfere with the behaviour of those particles, because the observation involves an exchange of energy between observer and the particle being observed. But you wouldn't be able to cook much in a quantum-sized pot.

YOU CAN FIND UNDERGROUND WATER BY DIVINING – Dowsing with bent coathangers, hickory sticks or divining rods, no matter how expensive the diviner or how impressive the credentials, cannot reveal the presence of underground water. There is no physical force that would move divining rods or any other device simply because of the presence of water beneath the earth below the diviner. Of course, scientific metal detectors, radar, remote satellite imaging and other geophysical techniques can and are used to detect subterranean features, including underground watercourses, oilfields and even lost cities.

PICKING DANDELIONS WILL MAKE YOU 'WET THE BED' – Dandelions do contain a mild diuretic, a natural chemical that increases the production of urine. However, simply picking the flowers will not lead to the ingestion of sufficient quantities of this chemical to cause uncontrolled nocturnal urination. Ironically, despite the origin of the English name for the plant being old French *dent-de-lion*, 'lion's tooth', the modern French for this species (*Taraxacum*) is *pissenlit* (or in the vernacular, *pisse au lit*) and the English folkname is 'piss-a-bed'.

THE AVERAGE WOMAN WILL INADVERTENTLY INGEST THREE KILOGRAMS OF LIPSTICK IN HER LIFE – The average content of a tube of lipstick weighs about three grams. To ingest three kilograms of lipstick the average woman would have to eat the entire contents of a thousand lipsticks. Assuming the lady uses lipstick from the age of 16 years to 76 years, that would mean a whole lipstick swallowed every two or three weeks.

SHAVING MAKES HAIR GROW BACK THICKER – Hair is dead tissue, so shaving can have no effect on the shape or size of the follicles from which the hair grows. If this deceived wisdom were true, bald men would shave their heads regularly in order to remedy their lack of cranial hirsuteness.

OSTRICHES BURY THEIR HEAD IN THE SAND WHEN FRIGHTENED – Ostriches bob their head from side to side to keep an eye on possible threats and run away if the threat is big enough. Thrusting your head underground is not the best way to carry out surveillance.

A SHOCK CAN TURN YOUR HAIR GREY OVERNIGHT – The colour (or lack of colour, in the case of grey or white) of individual hairs is fixed unless chemically treated. Individual hairs, as they grow, may lose pigment and become grey, but hair grows slowly and this change would not become apparent overnight. It is possible that on a head with some grey hairs the loss of coloured hair could happen relatively quickly, boosting the proportion of remaining hairs that are grey. But such hair loss is rare and, again, does not occur overnight.

YOU NEED TO CHEW EACH MOUTHFUL OF FOOD THIRTY-TWO TIMES – You only need to chew each mouthful of food enough for it to form a bolus, along with saliva, that can be swallowed comfortably. There is no natural law that demands any specific number of chews. Obviously, a morsel of tough steak will require more chews than a bowl of blancmange.

READING BY DIM LIGHT WILL PERMANENTLY DAMAGE YOUR EYESIGHT – Reading in dim light might cause eyestrain, make you feel tired, or cause your eyes to feel dry because you will blink less. However, it will not damage the focusing power of your eyes or the sensitivity of the retina.

IT'S COLD ENOUGH TO FREEZE THE BALLS OFF A BRASS MONKEY! – The deceived wisdom is that this well-known saying referred to a brass device, known as a monkey, that was used on sailing ships to prevent cannonballs from rolling around on deck. There never was such a device, and most sailing ships did not have cannons. A drop in temperature does cause metals to contract, but the idea of cannonballs becoming detached in cold weather is nonsense. The phrase is simply a crude alternative to 'cold enough to freeze the ears off a brass monkey'. No one knows where that came from.

AFTER FIVE YEARS OF USE, THE CONTENTS OF YOUR PILLOW AND DUVET WILL MOSTLY HAVE BEEN REPLACED BY HOUSE DUST MITE FAECES AND THEIR CARCASSES – This is simply not true, as chemical analysis will testify, and so would estimates of the mass of house dust mites that might accrue over such a time period. My theory? This particular piece of deceived wisdom was fabricated simply to help manufacturers sell more new pillows and duvets.

VERTIGO IS A FEAR OF HEIGHTS – Despite the visual assertions of the eponymous Hitchcock movie, vertigo is not a fear of heights. It is a form of dizziness in which sufferers experience a feeling of movement even when entirely stationary. It is usually caused by a problem with the balance centres of the inner ear, and can cause nausea and vomiting.

EVOLUTION IS JUST A THEORY – Evolution is certainly a theory, but use of the word 'just' is to dismiss decades of understanding, experiment and evidence that provide us with the most compelling explanation for the origin of the diversity of living things we see on our planet.

HAIR AND FINGERNAILS CONTINUE TO GROW AFTER YOU DIE – Hair and fingernails require there to be living cells in the hair follicles and in the bed of the fingernails if they are to grow. As with all the other cells in a corpse, these cells die when the person dies. For several days after death, the skin and tissues of a body dry out and shrink, so largely dead tissues such as hair and nails may appear to protrude more from the follicles and fingers, but this is not actual growth.

BIRDS EXPLODE IF THEY EAT RAW RICE – Rice grains will eventually swell if placed in water. Luckily for cooks, the process occurs quite rapidly in a pan of boiling water, but luckily for a bird or any other animal, not in the stomach. Moreover, guts and gizzards are not solid vessels and pipes, and there is plenty of room for expansion should an animal eat the desiccated food and find the grains expanding within.

PAINTING THE FORTH BRIDGE OR THE GOLDEN GATE BRIDGE IS A NEVER-ENDING TASK – You might imagine workers beginning at one end of a bridge with pots of paint and big brushes and working their way towards the other end, and just as they finish the job it must be started all over again.

However, bridges are never painted in such a methodical but altogether inefficient manner. Also, specific areas of any given bridge weather at different rates, so one component will inevitably need to be replaced several times for another component between necessary paint jobs. In addition, modern paints and coatings will last at least twenty years, and it certainly doesn't take two decades to paint a bridge, even from end to end.

AN ONION CUT IN HALF WILL ABSORB PAINT FUMES AFTER DECORATING – There is no chemical mechanism by which the vapours of the volatile organic solvents used in household paints that are present in the air while we are decorating will be absorbed to any great degree by the exposed surface of a sliced onion. Adequate ventilation is the only way to allow the vapours to disperse. The smell of the onion may mask the smell of the solvents, but only if your nose is particularly close to the onion, and then the discomfort caused by the onion vapours will bring a tear to your eye more quickly than the lingering smell of paint. Your call.

IT'S TOO COLD TO SNOW – Nowhere on earth can be too cold for snow. Meteorologists know very well that it can snow at almost any temperature below a certain freezing threshold. Theoretically, if it were 273°C below freezing, at absolute zero – the coldest anything could ever be – then it would not snow, but even in the deepest, darkest depths of the Antarctic winter it has not dropped below minus 90°C since records began. The Antarctic is certainly not at absolute zero; even deep space isn't that cold.

HEDGEHOGS RUN AWAY FROM CARS INSTEAD OF ROLLING UP – Although this was mentioned in a school exam paper in the 1990s, there is no evidence to suggest that hedgehogs have evolved to run from the path of speeding cars rather than resorting to their more familiar defence of rolling up into a prickly ball. Indeed, if some hedgehogs did opt to run, they would statistically increase their chances of being hit by a car unless they knew to run to the side, out of the path of the car. The numbers of hedgehogs with the genes that made them runners rather than rollers would then fall relative to those with the original spiky defence tactic.

YOUR TONGUE IS DIVIDED INTO AREAS FOR TASTING SWEET, SOUR, BITTER AND SALT – Although popular science books and school textbooks often show a 'map' of the tongue with the sweet taste buds at the tip and bitter at the back, and other regions demarcated for salt and sour, this is a misconception. The relative sensitivity of different regions varies insignificantly, a fact known to scientists since the work of Virginia Collings in 1974. All areas on the tongue can taste all the different types of taste, including the fifth taste known as 'umami', or 'deliciousness', knowledge of which emerged from research into Japanese cuisine. Scientists also now recognize that there is the related, but distinct, taste of 'fat'.

• • • • • • • • • • • • •

Find out more
http://www.everythingyouknowisalie.co.uk
http://www.snopes.com
http://dsc.discovery.com/tv/mythbusters
http://www.forteantimes.com/strangedays/mythbusters

"

The scientist is not a person who gives the right answers, he's one who asks the right questions

"

Claude Lévi-Strauss

Goal!

The Deceived Wisdom
The team that scores first
will generally win the match.

• • • • • • • • • • • •

Sports pundits commentating on football, ice hockey and other goal-based games often argue that scoring the first goal of the match is important and that doing so gives the scoring team the upper hand and an odds-on chance of winning the match. This notion is commonly rolled out during 'play-off' games, which are often played more defensively than a standard league or cup match. The problem with this claim is that in ice hockey, and football, the scores are usually quite low, unlike in say, rugby or cricket, so the final result can hinge on just a single goal rather than a wider range of points scored. This distorts the statistical evidence: of course the team that scores first in a 1–0 match is the team that wins, but that doesn't mean that it was obvious that they would win from the moment they scored – nor does it prove that in a high-scoring game, whichever team scored first would be statistically more likely to win.

Researchers in Canada wanted to kick this sporty piece of deceived wisdom into touch once and for all to show that scoring the first goal is not the big advantage it's claimed to be. They used the mathematics of probability to calculate the chances of the first-goal team winning at specific points through the course of a match based on the number of minutes left to play. They also counted extra time in their statistical analysis.

The team considered a match between The Statisticians and The Deviators that has t minutes to go once the first goal has been scored. They applied a mathematical formula known as a Poisson distribution to see how many goals are then likely to be scored during the remaining t minutes. The research team explains that if both teams are striving to win, then there is an equal chance of them scoring after that first goal. Each team's league position and scoring record from past matches are also taken into account in the formula.

The formula plays out as follows: from the kick-off, The Statisticians have a 50:50 chance of winning, as do The Deviators. If The Statisticians score within the first five minutes, their chances of winning go up because The Deviators now have to score to prevent The Statisticians from winning. If they score much later in the first half, their chances of winning go up even more, because there is less time for The Deviators to catch up. If there is just 15 minutes of play left, then The Statisticians are almost certain to win. It's simple statistics.

Of course, probability and statistics are notoriously difficult to pin down in real life: teams are rarely matched equally, and there are the weather, injuries and home advantage to take into consideration. It is best not to take statistical punditry at face value, whether you are following the fortunes of the Toronto Maple Leafs in ice hockey or the Magpies in football.

• • • • • • • • • • • •

The Science

A mathematical analysis of a hypothetical football match shows that the chances that the first team to score will win the match rise as time passes. But real-life probabilities are unpredictable, and there is never a dead cert in a match.

• • • • • • • • • • • •

Find out more

http://www.livescience.com/
3637-goal-hockey-raise-odds-winning.html

The hangover cure fail

The Deceived Wisdom

It's possible to cure a hangover, if you can just find the right prescription for you (they include a full cooked English breakfast, a strong coffee and some aspirin or the hair of the dog that bit you ... meaning another alcoholic drink).

. . . .

Hangover cures don't work. If you wake up the morning after the night before with a thumping headache, a sick feeling in your stomach and a mouth that feels like the bottom of a parrot (or the bottom of a parrot's cage, depending on what you were drinking) because of overindulgence in alcoholic beverages, then there is nothing you can do to accelerate the processing of the toxins. The only true 'cure' for a hangover is not to drink alcohol in the first place.

Apparently, hangovers cost Britain about £2 billion a year in lost earnings through sufferers taking unpaid sick leave. Researchers writing in the *British Medical Journal* trawled the medical databases and the internet looking for hangover cures, and contacted experts and manufacturers to ask whether there had been any randomized controlled trials of medical interventions that could prevent or treat hangovers. The research paper does not say whether the researchers had a personal vested interest in proving a hangover remedy effective.

Their results revealed eight trials that tested eight different agents: propranolol (a beta-blocking drug), tropisetron (a drug for nausea and vertigo), tolfenamic acid (a painkiller), sugar, the dietary supplements borage, artichoke and prickly pear, and a yeast-based preparation. The results from those trials showed no significant positive effects on hangover recovery. The paper concluded that there is no compelling evidence that any complementary or conventional remedy works.

Interestingly, ethanol, which is the main active ingredient in beer, wine, spirits and other alcoholic drinks, actually plays only a minor role in producing the thirst, headache, fatigue, nausea, sweating, tremor, remorse and anxiety that chatacterize the common hangover. A hangover is at its worst when almost all of the ethanol imbibed has been metabolized in the liver to acetaldehyde and that too has been cleared from the blood. It seems that other chemicals in the alcoholic beverages are to blame for the hangover. These include polyphenols, methanol and histamine, all of which can have noxious effects on the body if excessive amounts of alcohol are drunk. This might explain why highly coloured booze is more likely to cause a hangover than clear, pure vodka, for instance.

Alcohol, or more properly ethanol, exerts its effects by latching on to the so-called GABA-A receptors in the brain. This interferes with lots of brain processes, giving rise to both the pleasant and the unpleasant effects of drinking alcohol experienced by most people who partake. Researchers have been investigating whether it might be possible to create a

"

A real hangover is nothing to try out family remedies on. The only cure for a real hangover is death

"

Robert Benchley

chemical that interacts with these receptors in a limited way to give people a 'designer' experience of drinking alcohol without the noxious effects. Such efforts may be of interest in scientific research, but do we really need another substance of abuse on the market?

Regardless, during the party season we will continue to see hangover remedies in magazines, on websites and at the pharmacy counter. Some of them may contain nothing more than paracetamol and caffeine, while others will claim that a handful of jalapeno peppers or a full English breakfast is what is needed. But the bottom line, parrot or otherwise, is that to avoid a hangover you do not need the hair of the dog that bit you, but simply to temper your intake. At the party to end all parties, you might find that boogie rather than booze is the cure you have been looking for.

* * * * * * * * * * * *

The Science
There is no compelling evidence to show that any so-called 'hangover cure' actually works.

* * * * * * * * * * * *

Find out more

http://news.bbc.co.uk/1/hi/health/4552142.stm

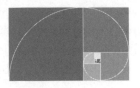

Yet another mixed bag of falsities

• • • • • • • • • • • •

HELIUM MAKES YOUR VOICE GO HIGH – You will hear all kinds of nonsense about helium being lighter than air and that this is what makes your vocal fold (commonly known as vocal chords) vibrate faster than normal and gives you the squeaky voice familiar to anyone who has sucked in the gas from a floating balloon. However, if you measure the main frequency at which the voice box is vibrating, you will find that it hasn't changed. Instead, the less dense helium simply emphasizes the higher-frequency harmonics of the note, so it sounds as if your voice is higher in pitch.

• • • • • • • • • • • •

Find out more

http://blog.sciencegeekgirl.com/2009/03/26/
myth-helium-makes-your-voice-high-pitched

THE GOLDEN RATIO HAS BEEN USED IN DESIGN FROM THE PARTHENON TO WIDESCREEN TVs – When widescreen TVs and computer monitors first became popular, there was a lot of talk about the ratio of the width of the screen to its height adhering to some ancient mathematics known to Greek philosophers and architects. The ratio of 1.618 to 1 (1.618 is, to three decimal places, one plus the square root of five, all divided by 2) was somehow 'golden' and was the most pleasing to the eye. It is approximated throughout art and in the widescreen TV format of 16 to 9 (though 16/9 is actually 1.778). If it was good enough for the Egyptians and their pyramids, or the dimensions of the Parthenon, then it is good enough for our TVs. Artists have apparently exploited the allegedly pleasing aesthetics of the golden ratio to make their paintings and photographs more beautiful.

The golden ratio is real. If a line is divided into two segments so that the ratio of the length of the whole line to the longer segment is the same as the ratio of the longer segment to the shorter one, then that ratio is the golden ratio. It can be seen in many natural contexts, such as the patterns formed by certain flowers. But according to the mathematician Keith Devlin, while it may be mathematically neat, it is not ubiquitous, it was not used to build the Parthenon, and it is not actually the ratio of most TVs and monitors – and when people are confronted with different rectangles, they do not consistently pick it out as being the most pleasing ratio.

• • • • • • • • • • • •

Find out more

http://www.lhup.edu/~dsimanek/pseudo/fibonacc.htm

http://www.maa.org/devlin/devlin_05_07.html

GOLDFISH HAVE A THIRTY-SECOND MEMORY – This particular piece of deceived wisdom has been debunked on several occasions by scientists and others who have trained goldfish to swim through mazes, push levers to obtain food and carry out other tasks. If their memory lasted just a few seconds, they simply wouldn't remember the route or what to do to get the food.

. . . .

Find out more

http://www.abc.net.au/science/articles/2010/01/14/2792407.htm

YOU CAN SEE THE GREAT WALL OF CHINA FROM THE MOON – You can see the Great Wall, motorways and even office blocks from low-earth orbit. Astronauts have taken photos to prove it. However, the moon is about 370,000 kilometres from the earth. It does not matter how long the Wall is – at that distance a structure as narrow as the Great Wall (a mere five metres at its widest) is well below the resolution of the human eye. The moon buggy left behind on the Apollo mission is only a little less than five metres long, but you can't see it when you're looking at the moon in the night sky. Even with the most powerful optical telescope, five metres is about ten times too small to discern from here on earth.

. . . .

Find out more

http://www.nasa.gov/vision/space/workinginspace/great_wall.html

THERE IS NO GRAVITY IN SPACE – This deceived wisdom presumably arose when people saw footage of astronauts in orbit floating weightlessly in their spaceship. However, an orbiting spaceship and the astronauts inside it experience the pull of the earth's gravity. It is only because they are in perpetual free fall towards the earth in their orbiting path that they look and feel weightless. Jump in a lift heading downwards, and you will experience the same feeling, however briefly.

Astronauts in earth orbit are only a tiny distance into 'space'. On much larger scales, gravity keeps the planets in orbit round the sun, and the stars in orbit round the centre of our galaxy. In fact, gravity pervades the whole universe: gravitational attraction gets weaker with increasing distance, but it never falls to zero.

. . . .

Find out more

http://www.astronomy.com/News-Observing/
Astronomy%20Myths/2009/11/
Theres%20no%20gravity%20in%20space.aspx

WATER GOES DOWN THE PLUGHOLE ANTICLOCKWISE IN THE NORTHERN HEMISPHERE – The Coriolis effect is a force caused by the rotation of the earth, which turns anticlockwise as you look down on it with the North Pole at the 'top'. This force is tiny, though over vast distances and over long time periods it can pull on weather systems. However, sinks and baths tend to be smaller than weather systems, and the water drains down their plugholes relatively quickly – too quickly and over too short a distance for the Coriolis effect to influence the direction of flow. Experiments have shown that the shape of the sink or bath, together with any initial swirling control, is what determines the direction of flow down the plughole, and that the flow can be in either direction in either hemisphere, regardless of what the signs tell tourists at the equator.

· · · · · · · · · · · · ·

Find out more

http://www.st-andrews.ac.uk/~dib2/climate/winds.html

YOUR FINGERS GO 'PRUNEY' AFTER A LONG BATH BECAUSE THEY ABSORB WATER – Until recently it was assumed that the reason your fingers and toes wrinkle up temporarily when you are in the bath or go swimming is that water is absorbed by the skin, and this causes the different skin layers to buckle. However, since 1935 scientists have known that the fingers of people with damage to the median nerve do not go completely pruney. Researcher Mark Changizi has put forward a theory that the wrinkled-finger effect is a reaction by the brain to our being wet which makes our fingers and feet wrinkle up to get a better grip on surfaces covered by the water, like the grip of a car's tyres in the wet.

· · · · · · · · · · · · ·

Find out more

http://changizi.wordpress.com/2011/07/07/
pruney-fingers-are-they-rain-treads/

A deathly deception

The Deceived Wisdom
The children's nursery rhyme 'Ring-a-ring o' Roses'
refers to the symptoms and fate of those who
succumbed to the Black Death in the
middle of the fourteenth century.

. . . .

The Black Death was an epidemic of bubonic plague, which is caused by the bacterium *Yersinia pestis*, carried by fleas carried by rats. It reached Europe via the Silk Road from Asia during the middle of the fourteenth century and is thought to have killed tens of millions of people across Europe.

The disease results in swellings on the skin from infected lymph glands. These swellings, called buboes, usually form in the neck, groin and armpits and leak pus and blood. These symptoms gave rise to the name 'bubonic plague'. The term Black Death refers to the gangrene that accompanies infection, which leads to blackened, dead tissue. The disease does not produce rings of 'roses' on the wrists. It cannot be treated with a 'pocket full of posies' or a 'bottle of posy', though one might imagine that people back then would carry scented flowers to mask the stench of death. The bacterium *Yersinia pestis* still exists and does respond to modern antibiotics if caught early enough. There are some concerns that a modern plague might emerge if outbreaks are not controlled or if the bacterium evolves resistance to the currently effective antibiotics.

The nursery rhyme itself was not published until the late nineteenth century, though there are mentions of a rhyme called 'Ring a Ring of Rosy' in the USA earlier than that. There are dozens of variants of this nursery rhyme, including ones about choosing a husband or the revelation of an illicit love affair. Some versions of the rhyme chant 'ashes, ashes' – alluding to death and burning (though most victims of the Black Death were actually buried in plague pits, rather than being cremated). Other versions say 'atishoo, atishoo' to represent the purportedly excessive sneezing of plague victims, while others simply say 'hush, hush', as if to keep a secret guarded:

Ring, a ring o' roses,
A pocket full o' posies,
Up-stairs and down-stairs,
In my lady's chamber –
Husher! Husher! Cuckoo!

For the nursery rhyme to have had its origins at the time of the Black Death it would need to have survived from generation to generation without being written down for several centuries in a repeatable and recognizable form, predating Chaucer's *Canterbury Tales*. However romantic it might be to imagine that a nursery rhyme about the Black Death could somehow persist for so long without being written down, there is no evidence that 'Ring-a-ring o' Roses' is anything but a relatively modern children's song.

. . . .

The Science

The symptoms of the Black Death (bubonic plague) are neither excessive sneezing nor 'rosy' patches on the skin, but buboes, hence the name. There is no evidence that children were singing the rhyme until at least the middle of the eighteenth century – and there are dozens of variations that seem to have many different allusions.

. . . .

Find out more

http://www.abc.net.au/science/articles/2008/07/08/2297794.htm

http://www.snopes.com/language/literary/rosie.asp

A final round of factitiousness

. . . .

Coffee does not help you sober up.

Climbing ivy does not cause walls to crumble.
If anything, it protects brickwork from weathering.

Drinking a fizzy drink and eating mints
will not burst your stomach.

Eating too many mints will not make men infertile.

Bulls are not angered by the colour red.

You cannot balance an egg on its end
at the equinoxes.

There is no dark side of the moon. There
is a side that is never visible from earth, but it is
nevertheless periodically illuminated by the sun.

You can teach an old dog new tricks.

The children's song 'Puff, the Magic Dragon'
is not about marijuana use.

. . . .

Afterword

. . . .

As you will hopefully have realized by now, much deceived wisdom is perfectly harmless, however misguided. At best it gives the pub know-alls and office bores the opportunity to show off their supposed knowledge. At worst, however, it can see people wasting their money on pointless products or risking their lives on fake healing and other misconceptions.

Deceived wisdom often has at its heart nothing more than a simple misunderstanding, and is propagated by a failure to fact-check. Much of the time it does nothing worse than persuade children to eat more carrots and crusts than they otherwise might and to not swallow chewing gum. It gives wine connoisseurs a seemingly endless topic of conversation and explains why cats and dogs (and their owners) are often at loggerheads.

But when deceived wisdom distracts those hoping to lose weight from following a sensible course of dietary action and an exercise plan, or leads those with serious diseases to fail to seek proper medical attention and instead to turn to the purveyors of sugar pills and placebos, then it can be a serious problem.

In this book I have tried to lay to rest many of the common factual fictions, the reality misconceptions, food fallacies and dietary deceptions. I hope that you will now know how to make the most of that bottle of wine without fuss, the reason you should continue to recycle your aluminium cans rather than send them to landfill, and why you should never accept

"

There are no forbidden questions in science, no matters too sensitive or delicate to be probed, no sacred truths

"

Carl Sagan

the urinary advances of a lifeguard, no matter how bad your jellyfish sting.

Here's one last piece of deceived wisdom for those of you curious enough to have made it this far: you have to be really clever to do science.

Why, thank you, but you flatter me. In fact, you don't have to have any special abilities to do science, just curiosity – something almost everyone is born with. We are inquisitive long before we can walk or even talk. As the very clever American scientist Neil deGrasse Tyson has put it, 'I can't think of any more human activity than conducting science experiments. Think about it – what do kids do? … They're turning over rocks, they're plucking petals off a rose – they're exploring their environment through experimentation.'

We are all scientists. We all have this innate curiosity that compels us to explore, to investigate, to find out more. It is a sad fact that our curiosity, that imaginative spark that wonders 'What if?' and generates more creativity than any artistic endeavour, is so often stifled by a dull teacher, a boring job or misconceptions about the world around us. It is at that point that science becomes something that other people do, the preserve of socially inept nerds in white lab coats and safety specs, a task fit only for those with wild hair and a distracted air. But we are all scientists; we are all born curious. Don't let the deceived wisdom of others distract you from that. Observe, experiment, understand – you owe that much to yourself as a member of that innately and insatiably curious species, *Homo sapiens*.

References

I have compiled a list of the links cited in each chapter on my Sciencebase website, http://sciencebase.com/dw-links. This is by no means an exhaustive list of the references I have used in researching and writing this book, but each link provides what I hope is a definitive validation of the debunking of the piece of deceived wisdom it relates to. Alternative versions of a select number of chapters have appeared on the Sciencebase website as well as in the 'Pivot Points' column in The *Euroscientist* online magazine.

Acknowledgements

Thanks to author and radio presenter Tim Lihoreau for helping to get the project started, to E&T's Olivia Bays for nurturing my first solo book so gently, and to John Woodruff for expert editing and for trapping escaped facts and oversights.

I am very grateful to the various scientific contacts who validated chapters, and to all those sceptics on the web who kindly provide information that corroborates facts in the book and tirelessly overturn the deceived wisdom. I am also indebted to my loyal Facebook and Twitter followers for their enthusiasm, suggestions and pre-orders, and to you for finding your own inner scientist.

Thanks to my children for their initial excitement when they learned that Dad was writing a book without co-authors, and for their amusement at the bald character on the cover, who bears no little resemblance to the author. Thanks too to my parents for supporting my inner geek.

Most of all, I want to thank my wife, Tricia, for her patience during the writing and editing process when late nights, early mornings and nocturnal 'light bulb moments' often precluded a good night's sleep for both of us. I hope you and she think it was all worth it.

About the author

David Bradley has worked in science communication for almost 25 years. He has written for *New Scientist*, the *Telegraph*, the *Guardian* and many other publications, as well as contributing to and editing books including *The Bedside Book of Chemistry*. He has won awards for his writing and blogging, including Daily Telegraph Science Writer of the Year. He lives in Cambridge, England, with his wife. He blogs at **www.sciencebase.com** and tweets as **@sciencebase** to more than 20,000 followers.

Let David know what you thought of *Deceived Wisdom*.

 Tweet him **@sciencebase**

www.sciencebase.com

Index